BUILDING SYSTEMS
REFERENCE GUIDE

The McGraw-Hill
Engineering Reference Guide Series

This series makes available to professionals and students a wide variety of engineering information and data available in McGraw-Hill's library of highly acclaimed books and publications. The books in the series are drawn directly from this vast resource of titles. Each one is either a condensation of a single title or a collection of sections culled from several titles. The Project Editors responsible for the books in the series are highly respected professionals in the engineering areas covered. Each Editor selected only the most relevant and current information available in the McGraw-Hill library, adding further details and commentary where necessary.

Hicks · CIVIL ENGINEERING CALCULATIONS REFERENCE GUIDE

Hicks · MACHINE DESIGN CALCULATIONS REFERENCE GUIDE

Hicks · PLUMBING DESIGN AND INSTALLATION REFERENCE GUIDE

Hicks · POWER GENERATION CALCULATIONS REFERENCE GUIDE

Hicks · POWER PLANT EVALUATION AND DESIGN REFERENCE GUIDE

Higgins · PRACTICAL CONSTRUCTION EQUIPMENT MAINTENANCE
     REFERENCE GUIDE

Johnson & Jasik · ANTENNA APPLICATIONS REFERENCE GUIDE

Markus and Weston · CLASSIC CIRCUITS REFERENCE GUIDE

Rosaler and Rice · INDUSTRIAL MAINTENANCE REFERENCE GUIDE

Rosaler and Rice · PLANT EQUIPMENT REFERENCE GUIDE

Merritt · CIVIL ENGINEERING REFERENCE GUIDE

Perry · BUILDING SYSTEMS REFERENCE GUIDE

Woodson · HUMAN FACTORS REFERENCE GUIDE FOR ELECTRONICS AND
     COMPUTER PROFESSIONALS

Woodson · HUMAN FACTORS REFERENCE GUIDE FOR PROCESS PLANTS

BUILDING SYSTEMS REFERENCE GUIDE

EDITOR-IN-CHIEF

ROBERT H. PERRY, Ph.D.

Consulting Chemical Engineer; Formerly
Project Engineer, Scientific Design Company;
Senior Technologist, Shell Chemical Corporation;
Process Engineer, Shell Development Company

Tyler G. Hicks, P.E.

Project Editor

McGraw-Hill Book Company

New York   St. Louis   San Francisco   Auckland   Bogotá
Hamburg   London   Madrid   Mexico   Milan
Montreal   New Delhi   Panama   Paris   São Paulo
Singapore   Sydney   Tokyo   Toronto

Library of Congress Cataloging-in-Publication Data

Building systems reference guide.

(McGraw-Hill engineering reference guide series)
"Drawn from the pages of Perry—Engineering manual,
published by McGraw-Hill Book Company"—Pref.
Includes index.
1. Structural engineering. 2. Buildings—Mechanical
equipment—Design and construction. 3. Buildings—
Environmental engineering. I. Perry, Robert H.,
1924–   . II. Hicks, Tyler Gregory, 1921–   .
III. Perry, Robert H., 1924–   . Engineering manual.
IV. Title. V. Series.
TA633.B85   1987   690   87-3172
ISBN 0-07-028802-X

1234567890   DOC/DOC   893210987

ISBN 0-07-028802-X

Building Systems Reference Guide reproduces Section 4 and a portion
of Section 7 of *Engineering Manual,* 3d ed., Robert H. Perry, editor,
McGraw-Hill, New York, 1976.

Printed and bound by R. R. Donnelley & Sons Company.

CONTENTS

Preface

Section 1. Building Systems Engineering

Structural   1-2
  Loads   1-2
  Structural Framing Systems   1-4
Mechanical   1-9
  Heating   1-9
  Moisture   1-26
  Cooling   1-29
  Air Distribution   1-68
  Refrigeration   1-76
  Water Distribution   1-77
  Pumps   1-78
  Drainage   1-79
  Cold Water   1-85
  Hot Water   1-88
  Gas Piping   1-91
Electrical   1-92
  Power Systems   1-92
  System Voltages   1-93
  Building Loads   1-96
  Distribution   1-99
  Motors and Controls   1-102
  Telephones   1-106
  Signal and Communications Systems   1-107
  Grounding   1-109

Section 2. Illumination

  Definition of Terms   2-2
  Inverse-square Law   2-2
  Application   2-4

Index   I-1

PREFACE

This is a concise building systems reference guide drawn from the pages of Perry—*Engineering Manual*, published by McGraw-Hill Book Company. It is a valuable reference for all working in building design, operation, and maintenance.

Topics covered in this useful guide include structural loads and framing systems; heating, moisture control, cooling, air distribution, refrigeration, water distribution, pumps, drainage, cold-water supply, hot-water supply, gas piping; power systems, voltages, building electrical loads, electrical distribution, motors and controls, telephones, signal and communication systems, grounding; illumination, levels for interior lighting, lamp data, coefficients of utilization, and brightness ratios.

Each topic is discussed in detail and backed by tabular and graphic data. A number of methods for making cost estimates of various building components are presented for use by the reader. Numerous standards and recommended conditions are given in concise, easy-to-access form.

Major users of this guide will be design engineers, architects, building operators, and building maintenance supervisors. For these users the guide will give quick access to vital information that is important in the design, operation, and maintenance of all types of modern buildings. Each of the contributors to this guide is an experienced building designer. That experience provides the reader with the key information needed to perform better on the job.

Tyler G. Hicks, P.E.

BUILDING SYSTEMS
REFERENCE GUIDE

SECTION 1

BUILDING SYSTEMS ENGINEERING

Gregory E. Brooks, B.C.E., M.C.E., P.E.; Chief Structural Engineer, Haines Lundberg & Waehler; Fellow, American Society of Civil Engineers; Member, American Concrete Institute, American Welding Society, Building Research Institute, National Society of Professional Engineers

Leander Economides, B.S.M.E., P.E., R.A.; Economides & Goldberg, Consulting Engineers; Member, American Society of Heating, Air Conditioning and Refrigeration Engineers, Building Research Institute, American Society of Mechanical Engineers, American Institute of Architects

Bronislaus F. Winckowski, B.E.E., P.E.; Chief Electrical Engineer, Haines Lundberg & Waehler; Senior Member, The Institute of Electrical and Electronics Engineers; Member, Illuminating Engineering Society, Building Research Institute, National Society of Professional Engineers

CONTENTS

Structural

1-1. Loads   1-2
1-2. Structural Framing
     Systems   1-4

Mechanical

1-3. Heating   1-9
1-4. Moisture   1-26
1-5. Cooling   1-29
1-6. Air Distribution   1-68
1-7. Refrigeration   1-76
1-8. Water Distribution   1-77
1-9. Pumps   1-78
1-10. Drainage   1-79

1-11. Cold Water   1-85
1-12. Hot Water   1-88
1-13. Gas Piping   1-91

Electrical

1-14. Power Systems   1-92
1-15. System Voltages   1-93
1-16. Building Loads   1-96
1-17. Distribution   1-99
1-18. Motors and Controls   1-102
1-19. Telephones   1-106
1-20. Signal and Communica-
      tions Systems   1-107
1-21. Grounding   1-109

STRUCTURAL

1-1. Loads

Structural Design. All structures are designed to sustain safely the weight of all permanent stationary construction (dead load) entering into a structure and the greatest loads induced by the intended occupancy (live load) or other uses.

Design Loads. The weights of materials most commonly used in building construction are tabulated and are classified as dead loads. Live loads are generally considered to be uniformly distributed and are classified according to occupancy.

Dead Load. Dead load is usually determined by the use of the weights given in Table 1-1.

Table 1-1

Building materials	Wt., lb/ft³	Building materials	Wt., lb/ft³
Aluminum	175	Fir	32
Ash, white	40	Granite, limestone, marble	165
Ashes, cinders	45	Iron, cast	450
Brass	534	Iron, wrought	485
Brick	120	Lead	710
Bronze	509	Maple	43
Cedar	22	Oak	59
Clay, dry	63	Paper	58
Clay, damp	110	Pine, white	26
Clay and gravel	100	Pine, yellow, long leaf	44
Concrete, lightweight aggregates	90	Poplar	30
Concrete, normal aggregates	150	Redwood	26
Concrete block, hollow, lightweight aggregates	65	Sand, gravel, dry, loose	90–105
		Sand, gravel, dry, packed	100–120
Concrete block, hollow, normal aggregates	85	Sandstone, bluestone	140
		Spruce	27
Copper	556	Steel	490
Earth, loose	76	Tin	459
Earth, packed	95	Zinc	440
Elm	45		

1-2

Table 1-1 (Continued)

Partitions and walls	Thick., in.	Wt., psf	Partitions and walls	Thick., in.	Wt., psf
Partitions:*			Interior walls (continued):		
Hollow plaster partition........	4	22	Concrete block, hollow-normal	12	97
Plaster on metal lath..........	¾	7	Concrete block, hollow-cinder	3	17
Steel studs—metal lath and				4	24
plaster (2 sides)............	4¾	15		6	33
Wood studs—wood lath				8	39
and plaster (2 sides)........	5⅜	14		12	63
Wood studs—metal lath			Gypsum block, solid.........	2	10
and plaster (2 sides)........	5¾	16		3	13
Wood studs—sheetrock........	4⅜	5			
			Exterior walls:*		
Interior walls:*			Brick......................	4	40
Brick......................	4	40		8	80
Clay tile, hollow block........	3	17		12	120
	4	18	Block, cinder, hollow........	8	38
	6	25		12	61
	8	31	Block, cinder, solid..........	8	48
Concrete block, hollow-normal..	3	26		12	72
	4	35	Block, normal aggregate......	8	59
	6	50		12	97
	8	59			

Floors, ceilings, & roofing	Thick., in.	Wt., psf	Floors, ceilings, & roofing	Thick., in.	Wt., psf
Floors:			Roofing (continued):		
Asphalt—mastic..............	1	12	Cooper sheet...............	...	2
Asphalt—tile................	⅛	1	Concrete plank.............	2	13
Cement or terr. finish........	1	13		2¾	18
Cinder concrete fill..........	2	10	Felt, 4 layers.............	...	1
Cinder concrete plank........	2	15	Foamglass (insulation).......	1	1
Concrete, lightweight........	1	8	Gypsum (fill)..............	1	3
Concrete, normal............	1	12	Gypsum slab, precast........	2	12
Floor plate.................	⅜	16		3	14
Grating (1-1¼ × ¼₆)........	1¼	9	Lead......................	⅛	8
Cellular metal flooring........	1½	5	3-ply roofing...............	...	1
	3	7	4-ply felt and gravel........	...	6
			5-ply felt and gravel........	...	7
Ceilings:			Sheathing..................	1	3
Insulation..................	1	2	Shingle, asbestos...........	...	4
Plaster on concrete..........	½	3	Shingle, wood..............	...	2
Plaster on metal lath........	¾	7	Slag roofing................	...	5
Plaster on suspended metal lath	...	10	Slate.....................	¼	10
Plaster on wood lath.........	⅞	6	Steel (No. 20 gauge)........	...	4
Pressed steel (No. 18 gauge)....	...	3	Tile, flat..................	...	18
Sheetrock..................	½	2	Tile, Spanish..............	...	8
			Tin.......................	...	1
Roofing:			Transite...................	...	4
Cinder fill.................	1	5	Wood roofers (av)..........	...	3
Concrete channel slab, light-					
weight, precast............	3½	14			

* Weights given do not include plaster on any surface. Add 5 psf for plaster applied directly on each face. Add 7 psf for metal lath and plaster applied on each face.

Live Load. Live loads generally used for buildings are shown in Table 1-2.

Table 1-2

Occupancy	Live load, psf	Occupancy	Live load, psf
Public buildings:		Hospitals, etc. (*Continued*):	
Armories..........................	150	Operating rooms.................	60
Auditoriums, churches, etc.:		Corridors, laboratories............	100
Fixed seats.......................	60	Residential buildings:	
Movable seats....................	100	Living areas......................	40
Exhibition buildings:		Corridors.........................	100
Restaurants, etc..................	100	Business buildings:	
Schools:		Office buildings...................	80
Classrooms........................	40	Light manufacturing...............	125
Corridors.........................	100	Heavy manufacturing..............	175 min
Other public buildings...............	80	Storage buildings:	
Theaters (stage floor)..............	150	Garages..........................	100
Institutional buildings:		Light warehouses.................	125
Hospitals, etc.:		Heavy warehouses.................	250 min
Private rooms and wards..........	40		

Wind Load. Vertical walls should be designed to resist a wind load, acting either inward or outward, as shown in Table 1-3.

Table 1-3

Height, ft	Wind press., psf	Height, ft	Wind press., psf
Less than 50.........................	20	100–199............................	28
50–99...............................	24	200 and above......................	30

Roof Wind and Snow Loads. Roofs should be designed for a combined wind and snow load of from 25 to 45 psf. The lesser value would be used for flat roofs in no-snow areas, and the greater values for flat roofs in heavy-snow areas. Intermediate values would be used for sloping roofs in all areas.

Other Loads. Other types of loading that should be considered in particular cases include earthquake loads, excessive wind loads, and impact loads.

1-2. Structural Framing Systems

The most economical structural framing system for a particular building is predicated on many variable conditions: use, locale, availability of material, fire resistivity, and magnitude of construction, to mention a few. The cost per square foot of floor area of an interior bay cannot be assumed as the average cost of the whole area, since the exterior-wall spandrels, elevator shafts, and wind-bracing and foundation conditions are not reflected in this single area, and these factors assume greater proportions of total building cost for smaller buildings than for larger buildings. Cost of various modes of design, however, can be determined by the design of an interior bay, which will indicate to some degree the most economical system to be chosen for a particular use and occupancy.

The unit prices used in this comparison of necessity reflect average prices for a certain locality. Any one unit price might be subject to variation. However, it is believed that the relationship between systems will not be materially affected. For the purpose of such a comparison, eight different designs of an interior bay are shown and described in Fig. 1-1 as schemes 1 to 8.

The designs are made for an office load of 80 + 20 psf, the 20 psf being an allowance for lightweight movable partitions, as required by most building codes. Weight in pounds per square foot includes floor fills, floor arches, beams, girders, and the average column weight for a 10-story structure, plus 85 per cent of the live load.

Indicated cost includes only the separately listed parts, and does not include contractor's profit, sales tax, and overhead charges.

The table of relative cost (Table 1-4) is so arranged to show:

Table 1-4. Construction Cost Analysis

Construction	Structural steel framing					Reinforced concrete		
	Scheme 1	Scheme 2	Scheme 3	Scheme 4	Scheme 5	Scheme 6	Scheme 7	Scheme 8
Live + dead load, psf*.......	160	163	182	203	150	192	222	208
Depth of construction........	1'11"	2'0¼"	2'2½"	2'2⅜"	2'7½"	2'1½"	2'1½"	1'2"
1 Concrete...............	0.58	0.54	0.95	0.91	0.47	1.38	1.63	1.30
2 Reinforcing steel........	0.18	......	0.37	0.44	0.07	1.65	1.38	1.70
3 Structural steel.........	2.38	2.61	2.57	2.99	1.87	......	......	......
4 Open web joist.........	......	......	......	......	1.47	......	......	......
5 Metal cellular deck......	1.12	2.28	......	......	......	......	......	......
6 Forms................	......	......	2.50	2.50	0.26	2.50	2.75	1.96
7 Monolithic fl. finish......	0.23	0.23	0.23	0.23	0.23	0.23	0.23	0.23
8 Subtotal..............	4.49	5.66	6.62	7.07	4.37	5.76	5.99	5.19
9 Cost ratio..............	100	126	147	157	97	128	133	116
10 Fire retardant..........	0.96	0.91	0.18	0.18	1.93	......	......	......
11 Subtotal..............	5.45	6.57	6.80	7.25	6.30	5.76	5.99	5.19
12 Cost ratio..............	100	121	125	133	116	106	110	95
13 Elect. service ducts......	1.03	0.70	4.62	4.62	4.62	4.62	4.62	4.62
14 Hung ceiling...........	1.49	1.49	1.49	1.49	1.49	1.49	1.49	1.49
15 Total.................	7.97	8.76	12.91	13.36	12.41	11.87	12.10	11.30
16 Cost ratio..............	100	110	162	168	156	149	152	142

* Load includes floor arches, beams, girders, and columns, plus 85 per cent of live load. Depth of construction is from finished floor to underside of F. P. of girders.

1. A complete structural floor capable of supporting the design loads, but not containing underfloor electrical ducts or hung ceiling and with no fire protection rating (lines 8 and 9).

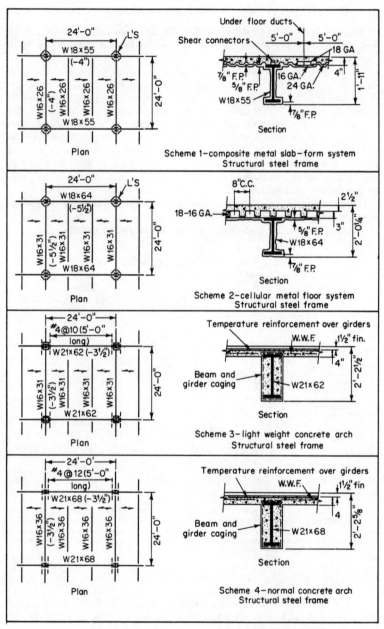

FIG. 1-1. Alternative construction schemes for interior bays.

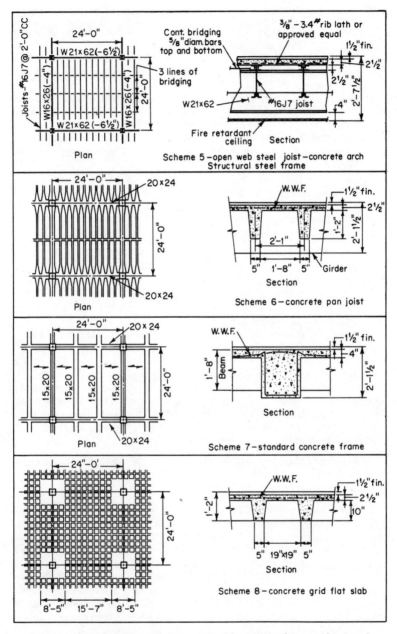

24'-0"
Joists—#16J7 @ 2'-0"CC
W16x26(-4")
W16x26(-4")
W21x62(-6½")
W21x62(-6½")
24'-0"

Plan

Cont. bridging
⅝" diam. bars top and bottom
3 lines of bridging
W21x62

³⁄₈"–3.4# rib lath or approved equal
1½"fin.
2½"
2½"
#16J7 joist
2'-7½"
4"
Fire retardant ceiling

Section

Scheme 5 — open web steel joist — concrete arch Structural steel frame

24'-0"
20x24
24'-0"
20x24

Plan

W.W.F.
1½"fin.
2½"
1'-2"
2'-1½"
5" 1'-8" 5"
2'-1"
Girder

Section

Scheme 6 — concrete pan joist

24'-0"
20 x 24
15x20 15x20 15x20 15x20
24'-0"
20x24

Plan

W.W.F.
1'-8"
Beam
1½" fin.
4"
2'-1½"

Section

Scheme 7 — standard concrete frame

24"-0'
24'-0"
8'-5" 15'-7" 8'-5"

Plan

W.W.F.
1½" fin.
2½"
10"
1'-2"
5" 19"x19" 5"

Section

Scheme 8 — concrete grid flat slab

FIG. 1-1. Alternative construction schemes for interior bays. (*Continued*)

1–7

2. Cost of adding required materials (if necessary) to give the maximum fire rating required by code (lines 11 and 12).

3. Cost for electrical flexibility, including additional ducts (if required) and cross headers. A finished ceiling is also included, but no mechanical duct work has been included.

Schemes 1 to 5 show structural steel framing of A36 steel, using a maximum unit stress of 24,000 psi, except for schemes 3 and 4, in which beam caging is used to allow a unit stress of 27,000 psi (AISC spec. sec. 1.11.2).

Schemes 6 to 8 are designs in reinforced concrete, using 3,000 psi concrete, with the exception of scheme 8, which requires 4,000 psi concrete to satisfy shear conditions.

For all eight floor systems without underfloor electrical ducts a steel-troweled monolithic cement floor finish has been included. In the case of schemes 3 to 8, the underfloor ducts are installed above the structural concrete floor slab, requiring a concrete fill of 1½ in. in depth to accommodate the ducts. This fill is charged to the cost of electric service ducts (line 13).

Fire Retardant Rating

A fire retardant rating of 3 hr is attained in the floor systems of all eight designs, with a 4-hr rating for column protection (shown on lines 10 to 12). Sprayed-on fire retardant is used in schemes 1 and 2.

Scheme 5, using open-web joist, is protected by a fire retardant ceiling attached to the underside of the joist and girders.

All columns of structural steel are covered with metal lath and fire retardant plaster or other approved fire protection giving a 4-hr rating.

The bar reinforcement or structural steel of schemes 3, 4, 6, 7, and 8 is protected by the minimum thickness of concrete cover required by code.

Scheme 1 is a design of composite construction, using shear connectors on both beams and girders, utilizing a light-gauge-steel form as reinforcement for the concrete slab. Where underfloor electrical ducts are required, cellular units are utilized. These units are usually placed 5 ft center to center.

The cost for electrification is estimated as follows:

$0.33 per ft² cellular units (differential increase in cost due cellular section)
  0.70 per ft² for cross-header ducts
—————
$1.03 per ft² (scheme 1, line 13)

Scheme 2 floor system uses metal cellular flooring with cells 8 in. center to center. No shear connectors are used in this type of floor; therefore the design is not composite in nature, requiring heavier beams and girders than are used in scheme 1. If underfloor electrical ducts are required, scheme 2 offers the greatest flexibility in the layout of underfloor ducts, having a cellular section every 8 in. The only extra cost involved for electrification of the floor system will be $0.70 per square foot for the cross-header ducts.

Schemes 3 and 4 are the usual beam and girder designs with concrete arches. Scheme 3 was included to show a comparison between lightweight and normal-weight concrete. An underfloor electrical-duct system for schemes 3 to 8 requires 1½ in. of concrete fill to accommodate the ducts.

The estimated cost for underfloor ducts will be:

$0.39 for concrete fill (the difference in unit cost)
4.23 for grid system of ducts (including headers)
$4.62 (schemes 3 to 8, line 13)

The details of these floor systems of construction are clearly shown in the plans and sections of their particular reference.

Table 1-4 clearly demonstrates that a given framing system can change its relative economic position, depending on the design criteria established for the particular building being analyzed.

Whereas scheme 5 is the most economical for a building requiring no fire protection, it drops to fifth place when fire protection is required. Scheme 8 at no extra cost now becomes the most economical. However, when electrification of the floor becomes a design criterion, scheme 1 or 2 becomes the economical choice.

MECHANICAL

This section includes design criteria, data, and physical laws and formulas applicable to the design of building heating, ventilating, air-conditioning, and plumbing systems.

1-3. Heating

Heating Load

$$Q_t = Q_{tr} + Q_{inf} + Q_{vent} \qquad (1\text{-}1)$$

where

Q_t = total heating load, Btu/hr

Q_{tr} = transmission load = $AU(t_i - t_o)$ Btu/hr   (1-2)

Q_{inf} = infiltration load = 1.08 (cfm) $(t_i - t_o)$ Btu/hr   (1-3)
       (Tables 1-10 to 1-12)

Q_{vent} = ventilation load = 1.08 (cfm) $(t_i - t_o)$ Btu/hr   (1-4)
       (Tables 1-13 to 1-14)

where

A = area through which heat flow occurs, ft^2
U = over-all heat transfer coefficient = $1/R_t$ Btu/(hr)(ft^2)(°F)   (1-5)
    (Tables 1-6 to 1-8)
t_o = outside-air design dry-bulb temperature, °F (Fig. 1-2)
t_i = inside design dry-bulb temperature, °F (Table 1-5)
cfm = cubic feet per minute, air
R_t = thermal resistance, °F/(Btu)(hr)(ft^2)

$$R_t = R_1 + R_2 + R_3 + \cdots + R_n \qquad (1\text{-}6)$$

where $R_1, R_2, \ldots,$ are the resistances to heat flow of the individual components of a composite construction (see Table 1-9 for values of R for concrete floors).

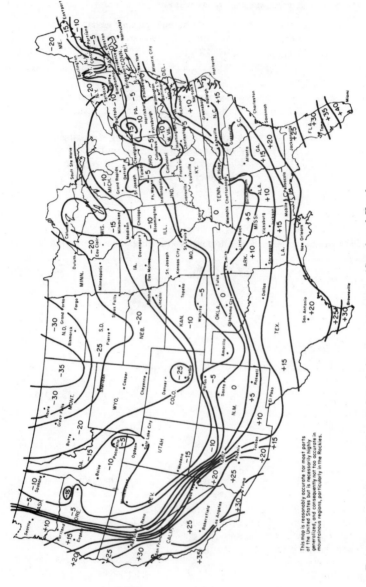

FIG. 1-2. Isotherms of winter outdoor design temperature. (*Strock and Koral,
"Handbook of Air Conditioning, Heating, and Ventilating," 2d ed., 1965.*)

This map is reasonably accurate for most parts
of the United States but is necessarily highly
generalized, and consequently not too accurate in
mountainous regions, particularly in the Rockies.

Table 1-5. Recommended Inside Design Conditions—Winter

Type of application	Winter				
	With humidification			Without humidification	
	Dry-bulb, F	Rel. hum., %	Temp. swing,* F	Dry-bulb, F	Temp. swing,* F
General Comfort Apartment, house, hotel, office, hospital, school, etc..........................	74–76	35–30	−3 to −4	75–77	−4
Retail Shops (Short-term occupancy) Bank, barber or beauty shop, depart-ment store, supermarket, etc........	72–74	35–30†	−3 to −4	73–75	−4
Low Sensible Heat Factor Applications (High latent load) Auditorium, church, bar, restaurant, kitchen, etc.........................	72–74	40–35	−2 to −3	74–76	−4
Factory Comfort Assembly areas, machining rooms etc..	68–72	35–30	−4 to −6	70–74	−6

SOURCE: "Carrier Corporation System Design Manual," Part I, Load Estimating, 1970.
* Temperature swing is below the thermostat setting at peak winter load conditions (no lights, people, or solar heat gain).
† Winter humidification in retail clothing shops is recommended to maintain the quality texture of goods.

Transmission Load

Table 1-6. Over-all Heat Transfer Coefficient *U*

Air-to-air heat transfer, Btu/(hr)(ft²)(°F)
Outside air 15-mph wind, inside still air

Example	Construction	*U*
Frame walls...............	Wood siding, building paper, air space, gypsum lath, plaster	0.24
	Wood siding, insulation board, air space, gypsum board	0.19
Frame partition............	Gypsum board, air space, gypsum board	0.34
Frame construction ceilings and floors	Linoleum or tile, felt, plywood, wood subfloor, air space, metal lath, plaster	0.23
Pitched roofs..............	Asphalt shingles, building paper, wood sheathing, air space, gypsum lath, plaster	0.28
Masonry wall..............	Face brick 4″, common brick 4″	0.48
	Face brick 4″, common brick 4″, air space, gypsum lath, plaster	0.29
	Face brick 4″, concrete block 4″, air space, gypsum lath, plaster	0.26
Masonry partition...........	Cement block (cinder aggregate), plaster on both sides	0.31
Concrete floor and ceiling....	Tile, felt, plywood ⅜″, air space, metal lath, plaster	0.23
Flat masonry roof..........	Built-up roofing, roof insulation 1″, concrete slab 4″, air space, metal lath, plaster	0.18

Table 1-7. Coefficient of Heat Transmission *U* for Windows and Skylights

Air-to-air heat transfer, Btu/(hr)(ft²)(°F)
Outside air 0°F, 15-mph wind, no solar radiation; inside still air

Construction	Vertical glass sheets		Horizontal glass sheets	
	Outdoor exposure	Indoor exposure	Outdoor exposure	Indoor exposure
Common window glass, single sheet................	1.13	0.75	1.22	0.96
Common window glass, two sheets, 1″ air space.......	0.53	0.45	0.63	0.56

Table 1-8. Coefficient of Heat Transmission *U* for Wood Doors

Air-to-air heat transfer, Btu/(hr)(ft²)(°F)
Outside air 0°F, 15-mph wind, no solar radiation; inside still air

Construction	Outdoor exposure	
	Single	With glass storm door
1″-thick solid door (²⁵⁄₃₂″)...............	0.54	0.37
2″-thick solid door (1⅝″)...............	.43	.28
Door containing wood or glass panels......	.85	.39

Table 1-9. Heat Loss of Concrete Floors at or Near Grade Level per Foot of Exposed Slab Edge, Btuh*

Outdoor Design Temperature, F	Total Width of Insulation, In.	Value of F for Unheated Slab[a]			Value of F for Heated Slab[b]		
		R = 5.0	R = 3.75	R = 2.50	R = 5.0	R = 3.33	R = 2.50
−30 and colder	24	34	51	67	46	69	92
−25 to −29	24	32	48	64	44	66	88
−20 to −24	24	30	45	60	41	61	82
−15 to −19	24	28	43	57	39	59	78
−10 to −14	24	27	40	51	37	55	74
− 5 to − 9	24	25	38	51	35	52	70
0 to − 4	24	24	36	48	32	48	64
+ 5 to + 1	24	22	33	44	30	45	60
+10 to + 6	18	21	31	42	25	38	50
+15 to +11	12	21	31	42	25	38	50
+20 to +16	Edge only	21	31	42	25	38	50

* Reprinted by permission from "ASHRAE Handbook of Fundamentals," ASHRAE, New York, 1972.
F = Heat loss coefficient, Btuh (per linear foot of exposed edge.)
R = Thermal resistance of insulation, $1/C$.
[a] Where perimeter insulation is not required, use $F = 50$ for unheated slabs or $F = 75$ for heated slabs.
[b] Slab floors having heating pipes or ducts under the slab shall be considered as *heated slabs.*

Table 1-10. Infiltration Through Double-Hung Wood Windows

Expressed in cubic feet per (hour) (foot of crack)

Type of window	Pressure difference (inches of water)				
	0.10	0.20	0.30	0.40	0.50
Wood double-hung window (Locked) (Leakage expressed as cubic feet per (hour) (foot of sash crack); only leakage around sash and through frame given)					
Nonweatherstripped, loose fit*	77	122	150	194	225
Nonweatherstripped, average fit†	27	43	57	69	80
Weatherstripped, loose fit	28	44	58	70	81
Weatherstripped, average fit	14	23	30	36	42
Frame-wall leakage‡ (Leakage is that passing between the frame of a wood double-hung window and the wall)					
Around frame in masonry wall, not caulked	17	26	34	41	48
Around frame in masonry wall, caulked	3	5	6	7	8
Around frame in wood frame wall	13	21	29	35	42

* A $\frac{3}{32}$-in. crack and clearance represent a poorly fitted window, much poorer than average.

† The fit of the average double-hung wood window was determined as $\frac{1}{16}$-in. crack and $\frac{3}{64}$-in. clearance by measurements on approximately 600 windows under heating season conditions.

‡ The values given for frame leakage are per foot of sash perimeter, as determined for double-hung wood windows. Some of the frame leakage in masonry walls originates in the brick wall itself, and cannot be prevented by caulking. For the additional reason that caulking is not done perfectly and deteriorates with time, it is considered advisable to choose the masonry frame leakage values for caulked frames as the average determined by the caulked and noncaulked tests.

Reprinted by permission from "ASHRAE Handbook of Fundamentals," ASHRAE, New York, 1972.

Table 1-11. Infiltration through Walls

Expressed in cubic feet per (hour) (square foot)

Type of wall	Pressure difference, inches of water				
	0.05	0.10	0.20	0.30	0.40
Brick Wall*					
8½ in. plain...................	5	9	16	24	28
plastered†...............	0.05	0.08	0.14	0.20	0.27
13 in. plain..................	5	8	14	20	24
plastered†...............	0.01	0.04	0.05	0.09	0.11
plastered‡...............	0.03	0.24	0.46	0.66	0.84
Frame wall, lath and plaster §.......	0.09	0.15	0.22	0.29	0.32

* Constructed of porous brick and lime mortar—workmanship poor.
† Two coats prepared gypsum plaster on brick.
‡ Furring, lath, and two coats prepared gypsum plaster on brick.
§ Wall construction: bevel siding painted or cedar shingles, sheathing, building paper, wood lath, and three coats gypsum plaster.
Reprinted by permission from "ASHRAE Handbook of Fundamentals," ASHRAE, New York, 1972.

Table 1-12. Infiltration through Doors—Winter*
15 Mph Wind Velocity†
Doors on One or Adjacent Windward Sides‡

Description	Cfm/ft² area§				
	Infrequent use	Average use			
		1- and 2-story building	Tall buildings, ft		
			50	100	200
Revolving door......................	1.6	10.5	12.6	14.2	17.3
Glass door (³⁄₁₆″ crack)...............	9.0	30.0	36.0	40.5	49.5
Wood door 3′ × 7′...................	2.0	13.0	15.5	17.5	21.5
Small factory door...................	1.5	3.0			
Garage and shipping-room door.........	4.0	9.0			
Ramp garage door...................	4.0	13.5			

* All values are based on the wind blowing directly at the window or door. When the prevailing wind direction is oblique to the window or doors, multiply the values by 0.60 and use the total window and door area on the windward side(s).
† Based on a wind velocity of 15 mph. For design wind velocities different from the base, multiply the table values by the ratio of velocities.
‡ Stack effect in tall buildings may also cause infiltration on the leeward side. To evaluate this, determine the equivalent velocity (V_e) and subtract the design velocity (V). The equivalent velocity is

$$V_e = \sqrt{V^2 - 1.75a} \text{ (upper section)}$$
$$= \sqrt{V^2 + 1.75b} \text{ (lower section)}$$

where a and b are the distances above and below the mid-height of the building, respectively, in feet.
Multiply the table values by the ratio $(V_e - V)/15$ for one-half of the windows and doors on the leeward side of the building. (Use values under one- and two-story building for doors on leeward side of tall buildings.)
§ Doors on opposite sides increase values 25 per cent.
SOURCE: "Carrier Corporation System Design Manual," Part I, Load Estimating, 1970.

Ventilation Load

Table 1-13. Minimum Outdoor Air Requirements to Remove Objectionable Body Odors under Laboratory Conditions

Type of occupants	Air space per person, ft³	Outdoor air supply, cfm per person
Heating season with or without recirculation. Air not conditioned.		
Sedentary adults of average socioeconomic status.......	100	25
	200	16
	300	12
	500	7
Laborers...	200	23
Grade school children of average socioeconomic status...	100	29
	200	21
	300	17
	500	11
Grade school children of lower socioeconomic status.....	200	38
Children attending private grade schools..............	100	22
Heating season. Air humidified by means of centrifugal humidifier. Water atomization rate 8 to 10 gph. Total air circulation 30 cfm per person.		
Sedentary adults..................................	200	12
Summer season. Air cooled and dehumidified by means of a spray dehumidifier. Spray water changed daily. Total air circulation 30 cfm per person.		
Sedentary adults..................................	200	<4

Reprinted by permission from "ASHRAE Handbook of Fundamentals," ASHRAE, New York, 1972.

Table 1-14. Outdoor Air Requirements

Application	Smoking	Cfm per person		Cfm/ft² of floor, min.
		Recommended	Min.	
Apartment:				
Average	Some	20	10	
Deluxe	Some	20	10	
Banking space	Occasional	10	7½	
Barber shops	Considerable	15	10	
Beauty parlors	Occasional	10	7½	
Brokers' board rooms	Very heavy	50	20	
Cocktail bars		40	25	
Corridors (supply or exhaust)				0.25
Department stores	None	7½	5	0.05
Directors' rooms	Extreme	50	30	
Drugstores	Considerable	10	7½	
Factories	None	10	7½	0.10
Five and ten cent stores	None	7½	5	
Funeral parlors	None	10	7½	
Garages		....	....	1.0
Hospitals:				
Operating rooms	None	....	....	2.0
Private rooms	None	30	25	0.33
Wards	None	20	10	
Hotel rooms	Heavy	30	25	0.33
Kitchens:				
Restaurant		....	....	4.0
Residence		....	....	2.0
Laboratories	Some	20	15	
Meeting rooms	Very heavy	50	30	1.25
Offices:				
General	Some	15	10	
Private	None	25	15	0.25
	Considerable	30	25	0.25
Restaurants:				
Cafeteria	Considerable	12	10	
Dining-room	Considerable	15	12	
Schoolrooms	None			
Shop, retail	None	10	7½	
Theater	None	7½	5	
	Some	15	10	
Toilets (exhaust)		....	....	2.0

Radiator and Convector Ratings. To determine the rating of a radiator or a convector for a given space, divide the heat loss of the space by the proper factor from Table 1-15 and select the radiator or convector having an equivalent Btu per hour rating. Thus, for a 75°F room with 1000 Btu heat loss, a convector fed with 1 psig steam should be selected for a 1110-Btu rating.

Table 1-15. Correction Factors for Various Types of Heating Units

Steam pressure (approx.) Gage vacuum, in. Hg	Psia	Temp. of heating medium steam or water	Cast-iron radiators Room temp., °F					Convectors Inlet air temp., °F					Finned tube Inlet air temp., °F					Baseboard Inlet air temp., °F				
			80	75	70	65	60	75	70	65	60	55	75	70	65	60	55	75	70	65	60	55
22.4	3.7	150	0.39	0.42	0.46	0.50	0.54	0.35	0.39	0.43	0.46	0.50	0.36	0.42	0.46	0.51	0.57	0.38	0.42	0.45	0.49	0.53
20.3	4.7	160	0.46	0.50	0.54	0.58	0.62	0.43	0.47	0.51	0.54	0.58	0.45	0.49	0.53	0.59	0.64	0.45	0.49	0.53	0.57	0.61
17.7	6.0	170	0.54	0.58	0.62	0.66	0.69	0.51	0.54	0.58	0.63	0.67	0.53	0.57	0.61	0.67	0.72	0.53	0.57	0.61	0.65	0.69
14.6	7.5	180	0.62	0.66	0.69	0.74	0.78	0.58	0.63	0.67	0.71	0.76	0.61	0.65	0.69	0.75	0.81	0.61	0.65	0.69	0.72	0.78
10.9	9.3	190	0.69	0.74	0.78	0.83	0.87	0.67	0.71	0.76	0.81	0.85	0.69	0.73	0.78	0.84	0.89	0.69	0.73	0.78	0.82	0.86
6.5	11.5	200	0.78*	0.83	0.87	0.91	0.95	0.76	0.81	0.85	0.90	0.95	0.77	0.81	0.86	0.92	0.97	0.81	0.86	0.92	0.95	1.00
Psig																						
1	15.6	215	0.91	0.95	1.00	1.04	1.09	0.90	0.95	1.00	1.05	1.10	0.91	0.94	1.00	1.06	1.11	0.91	0.95	1.00	1.05	1.09
6	21	230	1.04	1.09	1.14	1.18	1.23	1.05	1.10	1.15	1.20	1.26	1.03	1.08	1.14	1.19	1.24	1.04	1.09	1.14	1.19	1.25
15	30	250	1.23	1.28	1.32	1.37	1.43	1.27	1.32	1.37	1.43	1.47	1.20	1.26	1.31	1.37	1.43	1.22	1.27	1.32	1.37	1.43
27	42	270	1.43	1.47	1.52	1.56	1.61	1.47	1.54	1.59	1.67	1.72	1.38	1.44	1.50	1.56	1.62	1.43	1.47	1.52	1.59	1.64
52	67	300	1.72	1.75	1.82	1.89	1.92	1.85	1.89	1.96	2.04	2.08	1.67	1.73	1.79	1.86	1.92	1.75	1.82	1.89	1.92	1.96

SOURCE: Reprinted by permission from "ASHRAE Handbook and Product Directory," ASHRAE, New York, 1975.

Example of Use of Basic and Velocity Multiplier Charts
GIVEN:
Weight flow rate = 6,700 lb/hr
Initial steam pressure = 100 psig
Pressure drop = 11 psi/100 ft
FIND: Size of Schedule 40 pipe required and velocity of steam in pipe.
SOLUTION: The following steps are illustrated by the broken line in Fig. 1-3.
Step 1.   Enter diagram at a weight flow rate of 6,700 lb/hr and move vertically to the horizontal line at 100 psig.

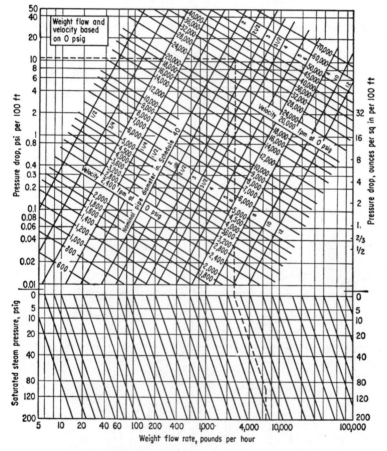

FIG. 1-3.   Basic chart for weight flow rate and velocity of steam in Schedule 40 pipe based on saturation pressure of 0 psig. (*Reprinted by permission from "ASHRAE Handbook of Fundamentals," ASHRAE, New York, 1972.*)

Step 2. Follow along inclined multiplier line (upward and to the left) to horizontal 0-psig line. The equivalent weight flow at 0 psig is about 2,500 lb/hr.

Step 3. Follow the 2,500-lb/hr line vertically until it intersects the horizontal line at 11 psi/100 ft pressure drop. The nominal pipe size is 2½ in. The equivalent steam velocity at 0 psig is about 32,700 fpm.

Step 4. To find the steam velocity at 100 psig, locate the value of 32,700 fpm on the ordinate of the velocity multiplier chart at 0 psig (Fig. 1-4).

Step 5. Move along the inclined multiplier line (downward and to the right) until it intersects the vertical 100-psig pressure line. The velocity as read from the right (or left) scale is about 13,000 fpm.

Note: The preceding steps 1 to 5 would be rearranged or reversed if different data were given.

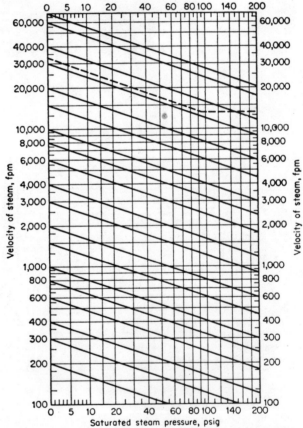

Fig.1-4. Velocity multiplier chart for use with Fig. 1–3. (*Reprinted by permission from "ASHRAE Handbook of Fundamentals," ASHRAE, New York*, 1972.)

Table 1-16. Weight Flow Rate of Steam in Schedule 40 Pipe[a] at Initial Saturation Pressures of 3.5 and 12 Psig[b,c]

Weight Flow Rate in Pounds per Hour

Pressure drop, psi per 100 ft in length

Nom. pipe size, in.	1/16 psi (1 oz)		1/8 psi (2 oz)		1/4 psi (4 oz)		1/2 psi (8 oz)		3/4 psi (12 oz)		1 psi		2 psi	
	\multicolumn{14}{c}{Saturation pressure, psig}
	3.5	12	3.5	12	3.5	12	3.5	12	3.5	12	3.5	12	3.5	12
3/4	9	11	14	16	20	24	29	35	36	43	42	50	60	73
1	17	21	26	31	37	46	54	66	68	82	81	95	114	137
1 1/4	36	45	53	66	78	96	111	138	140	170	162	200	232	280
1 1/2	56	70	84	100	120	147	174	210	218	260	246	304	360	430
2	108	134	162	194	234	285	336	410	420	510	480	590	710	850
2 1/2	174	215	258	310	378	460	540	660	680	820	780	950	1,150	1,370
3	318	380	465	550	660	810	960	1,160	1,190	1,430	1,380	1,670	1,950	2,400
3 1/2	462	550	670	800	990	1,218	1,410	1,700	1,740	2,100	2,000	2,420	2,950	3,450
4	640	800	950	1,160	1,410	1,690	1,980	2,400	2,450	3,000	2,880	3,460	4,200	4,900
5	1,200	1,430	1,680	2,100	2,440	3,000	3,570	4,250	4,380	5,250	5,100	6,100	7,500	8,600
6	1,920	2,300	2,820	3,350	3,960	4,850	5,700	7,000	7,200	8,600	8,400	10,000	11,900	14,200
8	3,900	4,800	5,570	7,000	8,100	10,000	11,400	14,300	14,500	17,700	16,500	20,500	24,000	29,500
10	7,200	8,800	10,200	12,600	15,000	18,200	21,000	26,000	26,200	32,000	30,000	37,000	42,700	52,000
12	11,400	13,700	16,500	19,500	23,400	28,400	33,000	40,000	41,000	49,500	48,000	57,500	67,800	81,000

[a] Based on Moody friction factor, where flow of condensate does not inhibit the flow of steam.

[b] The weight flow rates at 3.5 psig can be used to cover saturation pressure from 1 to 6 psig, and the rates at 12 psig can be used to cover saturation pressure from 8 to 16 psig with an error not exceeding 8 per cent.

[c] The steam velocities corresponding to the weight flow rates given in this table can be found from the basic chart and velocity multiplier chart, Fig. 1-3.

Reprinted by permission from "ASHRAE Handbook of Fundamentals," ASHRAE, New York, 1972.

Table 1-17. Return Main and Riser Capacities for Low-pressure Systems, Pounds per Hour

This table is based on pipe size data developed through the research investigations of The American Society of Heating, Refrigerating and Air-Conditioning Engineers.

	Pipe size, in.	1/32 psi or 1/2 oz drop per 100 ft			1/24 psi or 2/3 oz drop per 100 ft			1/16 psi or 1 oz drop per 100 ft			1/8 psi or 2 oz drop per 100 ft			1/4 psi or 4 oz drop per 100 ft			1/2 psi or 8 oz drop per 100 ft		
		Wet	Dry	Vac	Wet	Dry	Vac	Wet	Dry	Vac	Wet	Dry	Vac	Wet	Dry	Vac	Wet	Dry	Vac
	G	H	I	J	K	L	M	N	O	P	Q	R	S	T	U	V	W	X	Y
Mains	¾						42			100			142			200			283
	1	125	62		145	71	143	175	80	175	250	103	249	350	115	350			494
	1¼	213	130		248	149	244	300	168	300	425	217	426	600	241	600			848
	1½	338	206		393	236	388	475	265	475	675	340	674	950	378	950			1,340
	2	700	470		810	535	815	1,000	575	1,000	1,400	740	1,420	2,000	825	2,030			2,830
	2½	1,180	760		1,580	868	1,360	1,680	950	1,680	2,350	1,230	2,380	3,350	1,360	3,350			4,730
	3	1,880	1,460		2,130	1,560	2,180	2,680	1,750	2,680	3,750	2,250	3,800	5,350	2,500	5,350			7,560
	3½	2,750	1,970		3,300	2,200	3,250	4,000	2,500	4,000	5,500	3,230	5,680	8,000	3,580	8,000			11,300
	4	3,880	2,950		4,580	3,350	4,500	5,500	3,750	5,500	7,750	4,830	7,810	11,000	5,380	11,000			15,500
	5						7,880			9,680			13,700			19,400			27,300
	6						12,600			15,500			22,000			31,000			43,800
Risers	¾		48			48	143		48	175		48	249		48	350			494
	1		113			113	244		113	300		113	426		113	600			848
	1¼		248			248	388		248	475		248	674		248	950			1,340
	1½		375			375	815		375	1,000		375	1,420		375	2,000			2,830
	2		750			750	1,360		750	1,680		750	2,380		750	3,350			4,730
	2½						2,180			2,680			3,800			5,350			7,560
	3						3,250			4,000			5,680			8,000			11,300
	3½						4,480			5,500			7,810			11,000			15,500
	4						7,880			9,680			13,700			19,400			27,300
	5						12,600			15,500			22,000			31,000			43,800

Reprinted by permission from "ASHRAE Handbook of Fundamentals," ASHRAE, New York, 1972.

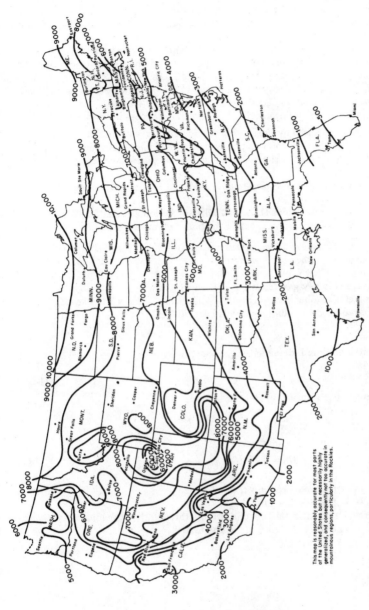

This map is reasonably accurate for most parts of the United States but is necessarily highly generalized, and consequently not too accurate in mountainous regions, particularly in the Rockies.

Fig. 1-5. Number of degree-days in a normal heating season. (*Strock and Koral, "Handbook of Air Conditioning, Heating, and Ventilating," The Industrial Press, New York, 1965.*)

Boiler Load. *Net Load* is the sum of direct-connected load components. These include direct radiation, infiltration, air tempering, humidification, hot water, process steam, and snow melting.

Design Load is the sum of the *net load* and the *piping tax*. Piping tax is the estimated heat emission in Btu per hour of the piping connecting the radiation and other apparatus to the boiler. In average heating systems it is common practice to consider the piping tax to be 20 per cent of the net load.

Gross, or Maximum, Load is the sum of the *design load* and the *pickup allowance*. Pickup allowance is the estimated increase in the normal load in Btu per hour, caused by the heating up of the cold system. For automatically fired boilers the sum of the piping tax and pickup allowance varies from 33.3 to 28.8 per cent of the *net load*. The larger percentage should be applied to smaller boilers.

Information on *boiler performance* and the *heating value of various fuels* may be found in Sec. 9.

Chimneys. Equations (1-7) to (1-10) are simplified equations for chimney sizes if the following typical values for boiler plants are assumed:

Average chimney gas temperature:	$T_c = 500°F$ (960°F abs)
Average atmospheric temperature:	$T_0 = 62°F$ (522°F abs)
Average coefficient of friction:	$f = 0.016$
Average chimney gas density at 0°F and 1 atm:	0.09 lb/ft^3
Barometer reading, sea level:	$B_0 = 29.92$ in. Hg

Required height of chimney above inlet, in feet:

$$H = 190D_r \tag{1-7}$$

Required minimum diameter of chimney, in feet:

$$d = 1.5W^{2/5} \tag{1-8}$$

Chimney gas velocity, in feet per second:

$$V_c = 13.7W^{1/5} \tag{1-9}$$

Stack draft, in inches of water:

$$D_r = 0.256HB_0 \left(\frac{1}{T_0 \text{ abs}} - \frac{1}{T_c \text{ abs}} \right) \tag{1-10}$$

where D_r = total required draft, in. of water
W = flue gas flow rate, lb/sec
Total required draft is the sum of draft loss through the breeching and through the boiler and the required draft in the firebox.

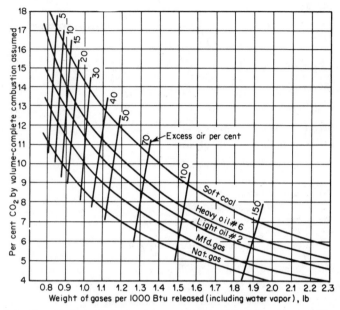

FIG. 1-6. Graphical evaluation and rate of flue gas flow from per cent CO_2 and fuel rate. *(Reprinted by permission from "ASHRAE Handbook & Product Directory," ASHRAE, New York, 1975.)*

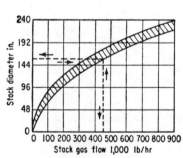

FIG. 1-7. Economical stack size based on approximately 5 per cent draft loss.

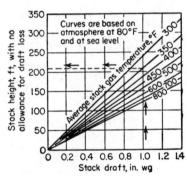

FIG. 1-8. Stack height as a function of stack draft.

1-4. Moisture

The moisture entering a building as water vapor may be expressed as

$$W_t = W_{trans} + W_{inf} + W_{vent} \qquad (1\text{-}11)$$

where W_t = total weight of vapor, grains

$$W_{trans} = \text{transmitted vapor} = MA\theta\,\Delta\rho \qquad \text{grains} \qquad (1\text{-}12)$$

$$W_{inf} = \text{air infiltrated vapor} = W(M_o - M_i) \qquad \text{grains} \qquad (1\text{-}13)$$

$$W_{vent} = \text{ventilation air vapor} = W(M_o - M_i) \qquad \text{grains} \qquad (1\text{-}14)$$

where M = permeance coefficient, perms

$$= \bar{\mu}/l,\ g/(ft^2)(hr)(Hg\,\Delta\rho) \quad (\text{Table 1-18}) \qquad (1\text{-}15)$$

A = area of flow path, ft^2

θ = time of transmission, hr

$\Delta\rho$ = vapor-pressure difference through flow path, **in. Hg**

W = weight of air, lb

M_o = moisture content of outside air, g/lb

M_i = moisture content of inside air, g/lb

where 1 perm = 1 $g/(ft^2)(hr)(in.\ Hg\ \Delta\rho)$

$\bar{\mu}$ = permeability, perm-in., $g\text{-}in./(ft^2)(hr)(Hg\,\Delta\rho)$

l = length of flow path, in.

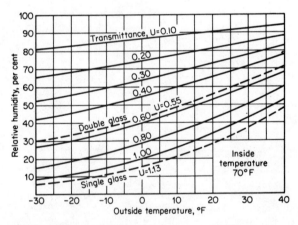

Fig. 1-9. Relative humidity at which visible condensation will appear on inside surface. (*Reprinted by permission from "ASHRAE Handbook of Fundamentals," ASHRAE, New York, 1972.*)

Table 1-18. Permeance and Permeability of Materials to Water Vapor[a]

Material	Permeance, perms			Permeability, perm-in.		
	Dry cup	Wet cup	Other	Dry cup	Wet cup	Other
Materials used in construction:						
Concrete (1:2:4 mix)	....	....	......	....	3.2	
Brick masonry (4 in. thick)...........	....	....	0.8			
Concrete block (8-in. cored, limestone aggregate).....	....	....	2.4			
Tile masonry, glazed (4 in. thick).....	....	....	0.12			
Asbestos-cement board (0.2 in. thick)..	0.54					
Plaster on metal lath (¾ in.)........	....	....	15			
Plaster on wood lath................	....	11				
Plaster on plain gypsum lath (with studs)...................	....	....	20			
Gypsum wallboard (⅜ in. plain)......	....	....	50			
Gypsum sheathing (½-in. asphalt impreg)......................	....	....		20		
Structural insulating board, sheathing quality...................	....	....		....	...	20–50
interior, uncoated, ½ in.............	....	....	50–90			
Hardboard, ⅛ in. standard..........	....	....	11			
⅛ in. tempered............	....	....	5			
Built-up roofing (hot-mopped)........	0.0					
Wood, sugar pine..................	....	....		....	...	0.4–5.4[b]
Plywood, douglas-fir, exterior glue, ¼ in. thick...........	....	....	0.7			
Plywood, douglas-fir, interior, glue, ¼ in. thick...................	....	....	1.9			
Acrylic, glass-fiber-reinforced sheet, 56 mil........................	0.12					
Polyester, glass-fiber-reinforced sheet, 48 mil........................	0.05					
Thermal insulations:						
Air (still).........................	....	....	....	....	...	120
Cellular glass.....................	....	....	....	0.0		
Corkboard........................	....	....	....	2.1–2.6	9.5	
Mineral wool (unprotected)..........	....	....	....	....	116	
Expanded polyurethane (R-11 blown) board stock..............	....	....	....	0.4–1.6		
Expanded polystyrene, extruded......	....	....	....	1.2		
bead..........	....	....	....	2.0–5.8		
Unicellular synthetic flexible rubber foam...........................	....	....	....	0.02–0.15		
Plastic and metal foils and films[c]:						
Aluminum foil, 1 mil..............	0.0					
0.35 mil..............	0.05					
Polyethylene, 2 mil.................	0.16					
4 mil.................	0.08					
6 mil.................	0.06					
8 mil.................	0.04					
10 mil.................	0.03					
Polyester, 1 mil..................	0.7					
Cellulose acetate, 125 mil...........	0.4					
Polyvinylchloride, unplasticized, 2 mil	0.68					
plasticized, 4 mil...	0.8–1.4					
Building papers, felts, roofing papers[d]:						
Duplex sheet, asphalt laminated, aluminum foil one side (43)[e]	0.002	0.176				
Saturated and coated roll roofing (326)[e]	0.05	0.24				
Kraft paper and asphalt laminated, reinforced 30-120-30 (34)[e]..	0.3	1.8				
Blanket thermal insulation back-up paper, asphalt coated (31)[e].........	0.4	0.6–4.2				
Asphalt-saturated and coated vapor-barrier paper (43)[e]...............	0.2–0.3	0.6				
Asphalt-saturated but not coated sheathing paper (22)[e].............	3.3	20.2				
15-lb asphalt felt (70)[e]..............	1.0	5.6				
15-lb tar felt (70)[e]..................	4.0	18.2				
Single-kraft, double infused (16)[e]......	31	42				

Table 1-18. Permeance and Permeability of Materials to Water Vapor (Continued)

Material	Permeance, perms			Permeability, perm-in.		
	Dry cup	Wet cup	Other	Dry cup	Wet cup	Other
Liquid-applied coating materials:						
Paint—2 coats:						
Asphalt paint on plywood	......	0.4				
Aluminum varnish on wood	0.3–0.5					
Enamels on smooth plaster	......	....	0.5–1.5			
Primers and sealers on interior insulation board	......	......	0.9–2.1			
Various primers plus 1 coat flat oil paint on plaster	......	......	1.6–3.0			
Flat paint on interior insulation board	......	......	4			
Water emulsion on interior insulation board	......	......	30–85			
Paint—exterior, 3 coats:						
White lead and oil on wood siding	0.3–1.0					
White lead-zinc oxide and oil on wood	0.9					
Styrene-butadiene latex coating, 2 oz/ft²	11					
Polyvinyl acetate latex coating, 4 oz/ft²	5.5					
Chloro-sulfonated polyethylene mastic,						
3.5 oz/ft²	1.7					
7.0 oz/ft²	0.06					
Asphalt cut-back mastic,						
⅟₁₆ in. dry	0.14					
³⁄₁₆ in. dry	0.0					
Hot melt asphalt,						
2 oz/ft²	0.5					
3.5 oz/ft²	0.1					

[a] Table 1-18 gives the water vapor transmission rates of some representative materials. Data are provided to permit comparisons of materials; but in the selection of vapor barrier materials, exact values for permeance or permeability should be obtained from the manufacturer of the materials under consideration or secured as a result of laboratory tests. A range of values shown in the table indicated variations among mean values for materials that are similar but of different density, orientation, lot or source. The values are intended for design guidance and should not be used as design or specification data. The compilation is from a number of sources; values from dry-cup and wet-cup methods were usually obtained from investigations using ASTM E96 and C355; values shown under *Other* were obtained from investigations using such techniques as *two-temperature, special cell, and air-velocity.*

[b] Depending on construction and direction of vapor flow.

[c] Usually installed as vapor barriers, although sometimes used as exterior finish and elsewhere near cold side where special considerations are then required for warm-side barrier effectiveness.

[d] Low permeance sheets used as vapor barriers. High permeance used elsewhere in construction.

[e] Basis weight in lb/500 ft².

Reprinted by permission from "ASHRAE Handbook of Fundamentals," ASHRAE, 1972.

Table 1-19. Grains of Moisture per Pound of Dry Air vs. Dew-point Temperature, °F

DP	Grains	DP	Grains	DP	Grains	DP	Grains	DP	Grains
0	5.50	16	12.36	32	26.40	48	49.50	64	89.18
1	5.79	17	12.99	33	27.52	49	51.42	65	92.40
2	6.10	18	13.63	34	28.66	50	53.38	66	95.76
3	6.43	19	14.30	35	29.83	51	55.45	67	99.19
4	6.77	20	15.01	36	31.07	52	57.58	68	102.8
5	7.12	21	15.75	37	32.33	53	59.74	69	106.4
6	7.50	22	16.53	38	33.62	54	61.99	70	110.2
7	7.89	23	17.33	39	34.97	55	69.34	71	114.2
8	8.30	24	18.17	40	36.36	56	66.75	72	118.2
9	8.73	25	19.05	41	37.80	57	69.23	73	122.4
10	9.18	26	19.9l	42	39.31	58	71.82	74	126.6
11	9.65	27	20.94	43	40.88	59	74.48	75	131.1
12	10.15	28	21.93	44	42.48	60	77.21	76	135.7
13	10.66	29	22.99	45	44.14	61	80.08	77	140.4
14	11.20	30	24.07	46	45.87	62	83.02	78	145.3
15	11.77	31	25.21	47	47.66	63	86.03	79	150.3

Table 1-19A. Vapor Pressure of Saturated Air, Inches of Hg, vs. Dry-bulb Temperature, °F

T	P_{sat}	T	P_{sat}	T	P_{sat}	T	P_{sat}	T	P_{sat}
0	.03764	16	.08461	32	.18035	48	.33629	64	.60073
1	.03966	17	.08884	33	.18778	49	.34913	65	.62209
2	.04178	18	.09326	34	.19546	50	.36240	66	.64411
3	.04400	19	.09789	35	.20342	51	.37611	67	.66681
4	.04633	20	.10272	36	.21166	52	.39028	68	.69019
5	.04877	21	.10777	37	.22020	53	.40492	69	.71430
6	.05133	22	.11305	38	.22904	54	.42004	70	.7391ᵇ
7	.05402	23	.11856	39	.23819	55	.43565	71	.76475
8	.05683	24	.12431	40	.24767	56	.45176	72	.79112
9	.05977	25	.13032	41	.25748	57	.46480	73	.81828
10	.06285	26	.13659	42	.26763	58	.48558	74	.84624
11	.06608	27	.14313	43	.27813	59	.50330	75	.87504
12	.06946	28	.14966	44	.28889	60	.52159	76	.90470
13	.07299	29	.15707	45	.30023	61	.54047	77	.93523
14	.07669	30	.16452	46	.31185	62	.55994	78	.96665
15	.08056	31	.17227	47	.32386	63	.58002	79	.99899

SOURCE: "ASHRAE Guide and Data Book," chap. 3, table 2, ASHRAE, New York, 1963.

1-5. Cooling

Cooling Load. Q_t, the total simultaneous cooling load, Btu/hr.

$$Q_t = Q_{ext} + Q_{int} + Q_{outside\ air} \tag{1-16}$$

where

$$Q_{ext} = \text{external heat gains} = Q_{transmission} + Q_{solar} \tag{1-17}$$

$$Q_{int} = \text{internal heat gains} = Q_{lights} + Q_{people}$$
$$+ Q_{equipment} + Q_{transmission} \tag{1-18}$$

$$Q_{tr\text{-}sol} = AU(sa\ \Delta t) \quad \text{for walls and roofs, Btu/hr} \tag{1-19}$$

where $sa\ \Delta t$ = sol-air equivalent temperature differential, °F

$$Q_{\text{glass}} = Q_{\text{solar}} + Q_{\text{tr}} \tag{1-20}$$

$$= \text{SHGF (SF)} A_1 + A_2 U(t_o - t_i) \tag{1-21}$$

where SHGF = solar heat gain factor (Tables 1-24 to 1-28)
        SF = shading factor (Tables 1-29 to 1-33)
        A_1 = area of sunlit glass, ft^2
        A_2 = area of total glass,
        U = over-all coefficient (Tables 1-22 and 1-23)
        t_o = outside design temp, °F (Fig. 1-12)
        t_i = inside design temp, °F (Table 1-40)

$$Q_{\text{lights}} = 3.41 \times \text{wattage input to conditioned space, Btu/hr} \tag{1-22}$$
(Figs. 1-10 and 1-11)

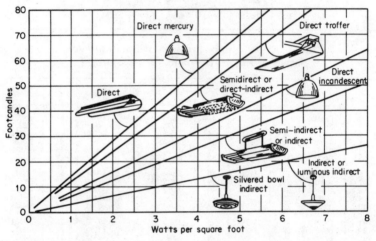

Fig. 1-10. Typical heat-gain lighting fixtures. On the curve above, follow the horizontal line, beginning at the maintained foot-candle value selected in (1), until it intersects the curve corresponding to the fixture type to be installed.

$$Q_{\text{people}} = \text{number of people } (q_{\text{sensible}} + q_{\text{latent}}), \text{ Btu/hr} \tag{1-23}$$
(Table 1-35)

$$Q_{\text{equipment}} = q_{\text{sensible}} + q_{\text{latent}} \qquad \text{Btu/hr (Tables 1-36 to 1-39)} \tag{1-24}$$

$$Q_{\text{outside air}} = q_{\text{sensible}} + q_{\text{latent}} \qquad \text{Btu/hr}$$
$$= 1.08(t_o - t_i) \text{ cfm} + 0.68(M_o - M_i) \text{ cfm} \qquad \text{Btu/hr} \tag{1-25}$$

where M_o = outside moisture content at design wet bulb °F, g/lb
        M_i = inside moisture content at design relative humidity, g/lb
        $Q_{\text{tr}} = AU(t_o - t_i)$ for partitions, floors, ceilings, Btu/hr $\qquad$ (1-26)

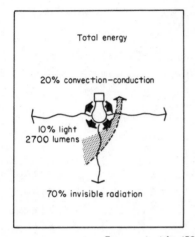

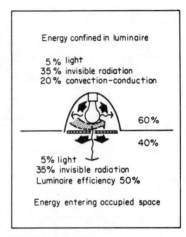

Energy output for 150 -watt Incandescent lamp

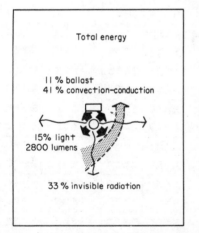

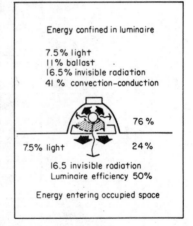

Energy output for 40 -watt Fluorescent lamp and ballast

FIG. 1-11. Distribution of energy output of incandescent and fluorescent lamps. (*Light Magazine*, vol. 29, fig. 1.)

Table 1-20A. Total Equivalent Temperature Differentials for Calculating Heat Gain through Sunlit Walls

Sun Time — Exterior color of wall: D = dark, L = light

Columns 8, 10, 12 = A.M.; columns 2, 4, 6, 8, 10, 12 = P.M. Each time column is split into D (dark) and L (light). λ = Amplitude Decrement Factor; δ = Time Lag, hr.

Group A (λ = 0.34, δ = 2 hr)

North Lat. Facing	8 D	8 L	10 D	10 L	12 D	12 L	2 D	2 L	4 D	4 L	6 D	6 L	8 D	8 L	10 D	10 L	12 D	12 L	South Lat. Facing
NE	27	16	31	18	26	17	24	17	24	18	23	17	20	15	17	13	15	11	SE
E	32	18	41	24	37	22	29	20	28	20	26	19	23	16	20	14	18	13	E
SE	25	15	36	21	38	23	33	23	28	21	26	18	22	16	19	14	18	12	NE
S	14	9	20	13	28	18	33	22	21	21	25	18	20	15	17	13	15	11	N
SW	17	11	20	13	24	16	22	16	42	27	41	26	28	19	20	14	18	12	NW
W	17	11	20	13	24	16	20	16	42	27	48	30	33	22	22	15	19	13	W
NW	14	9	17	11	21	14	23	14	31	17	38	25	28	19	18	13	16	11	SW
N	14	9	15	10	17	12	15	12	21	16	21	16	20	14	14	11	12	9	S

Group B (λ = 0.51, δ = 3 hr)

North Lat. Facing	8 D	8 L	10 D	10 L	12 D	12 L	2 D	2 L	4 D	4 L	6 D	6 L	8 D	8 L	10 D	10 L	12 D	12 L	South Lat. Facing
NE	12	7	27	14	31	17	30	19	31	21	30	22	27	20	21	17	16	13	SE
E	14	8	34	18	45	24	43	25	39	24	35	25	30	22	23	18	17	14	E
SE	9	5	25	13	39	21	44	26	41	26	37	25	31	23	24	18	17	14	NE
S	4	3	7	4	18	11	32	19	41	26	39	27	33	24	25	18	18	15	N
SW	5	3	7	4	11	7	23	15	41	26	54	34	51	33	38	25	26	19	NW
W	6	4	7	4	11	7	18	12	35	23	55	34	59	37	43	28	30	20	W
NW	5	3	6	4	11	7	17	12	26	18	41	27	47	31	36	24	25	18	SW
N	6	4	9	5	12	8	18	12	22	17	25	20	27	21	22	17	16	14	S

Group C (λ = 0.40, δ = 4 hr)

North Lat. Facing	8 D	8 L	10 D	10 L	12 D	12 L	2 D	2 L	4 D	4 L	6 D	6 L	8 D	8 L	10 D	10 L	12 D	12 L	South Lat. Facing
NE	9	6	19	10	26	15	28	17	29	18	29	20	28	20	24	19	20	16	SE
E	10	7	22	12	36	19	40	23	39	23	36	24	33	23	28	20	22	17	E
SE	8	6	16	9	29	16	38	21	39	24	37	24	34	24	28	21	23	17	NE
S	7	5	7	4	12	7	22	14	32	20	36	24	34	24	29	21	23	17	N
SW	9	6	8	5	10	6	16	9	28	18	42	26	48	30	42	28	33	22	NW
W	10	7	8	5	10	6	14	9	25	16	40	25	52	32	47	30	37	24	W
NW	8	6	8	5	9	6	13	9	19	14	30	20	40	27	38	26	30	21	SW
N	7	5	8	5	10	7	14	9	18	13	22	16	25	19	23	18	19	16	S

Group D^a

	SE	E	NE	N		NW	W	SW	S
NE	8	19	5	10		15	29	17	30
E	9	23	6	12		20	42	24	40
SE	7	16	5	9		16	40	22	41
S	5	6	4	4		7	23	14	34
SW	8	7	5	4		6	16	10	30
W	8	7	6	5		6	14	9	40
NW	7	6	5	4		6	13	9	38
N	6	8	4	5		6	14	10	38

0.45   4

Right labels: SE, E, NE, N / NW, W, SW, S

Group E^a

	SE	E	NE	N		NW	W	SW	S
NE	10	23	6	16		16	30	18	30
E	11	28	6	22		22	43	24	36
SE	8	20	5	15		19	42	24	38
S	4	6	3	11		9	28	17	39
SW	6	7	4	9		6	19	22	35
W	7	7	4	10		6	16	20	30
NW	6	6	4	10		6	15	16	26
N	6	8	5	11		7	16	15	24

0.48   4

Right labels: SE, E, NE, N / NW, W, SW, S

Group F^a

	SE	E	NE	N		NW	W	SW	S
NE	9	14	6	21		17	25	22	29
E	10	17	6	28		22	35	19	37
SE	9	13	6	22		21	31	16	37
S	9	7	5	10		16	17	11	32
SW	12	10	6	9		11	13	11	33
W	14	11	7	10		11	12	11	24
NW	12	9	6	9		11	15	14	19
N	8	8	6	9		11	12	15	21

0.32   6

Right labels: SE, E, NE, N / NW, W, SW, S

1-33

Table 1-20A. Total Equivalent Temperature Differentials for Calculating Heat Gain through Sunlit Walls (Continued)

Exterior color of wall—D = dark, L = light

Group Gᵃ (λ = 0.25, δ = 6 hr)

North Lat. Wall Facing	8 A.M. D	8 L	10 A.M. D	10 L	12 D	12 L	2 P.M. D	2 L	4 D	4 L	6 D	6 L	8 D	8 L	10 D	10 L	12 D	12 L	South Lat. Wall Facing
NE	11	9	10	15	20	12	24	14	25	16	26	17	27	18	26	18	23	17	SE
E	13	9	17	11	26	15	32	18	34	20	34	21	33	22	31	21	27	19	E
SE	13	9	14	9	21	12	28	16	33	16	34	21	33	22	31	21	27	19	NE
S	12	9	10	7	11	8	16	10	23	15	29	18	30	20	29	20	26	19	N
SW	16	11	13	9	13	8	14	9	20	13	29	19	37	24	39	25	35	23	NW
W	18	12	15	10	14	9	14	9	18	12	27	18	38	24	42	26	38	25	W
NW	14	10	12	8	12	8	15	9	16	11	24	15	33	20	33	22	31	21	SW
N	10	8	10	7	10	7	12	8	15	10	18	13	20	15	21	16	20	16	S

Group Hᵇ (λ = 0.14, δ = 8 hr)

North Lat. Wall Facing	8 A.M. D	8 L	10 A.M. D	10 L	12 D	12 L	2 P.M. D	2 L	4 D	4 L	6 D	6 L	8 D	8 L	10 D	10 L	12 D	12 L	South Lat. Wall Facing
NE	15	11	16	11	18	12	20	13	22	14	24	15	25	16	25	17	24	17	SE
E	18	13	18	12	22	14	26	16	29	17	30	19	31	20	30	20	29	19	E
SE	18	13	17	11	19	13	23	14	27	16	30	18	30	19	30	18	28	19	NE
S	16	12	14	10	14	10	15	10	19	12	23	15	26	17	26	18	26	18	N
SW	22	14	19	12	17	11	18	11	18	12	23	15	29	18	32	21	32	21	NW
W	23	15	20	13	18	12	18	11	18	12	22	15	29	18	33	21	34	22	W
NW	19	13	17	11	15	10	15	10	15	11	20	12	23	15	26	18	27	18	SW
N	13	10	12	9	11	9	13	9	13	10	15	11	17	13	18	14	19	14	S

Group Iᵇ (λ = 0.13, δ = 6 hr)

North Lat. Wall Facing	8 A.M. D	8 L	10 A.M. D	10 L	12 D	12 L	2 P.M. D	2 L	4 D	4 L	6 D	6 L	8 D	8 L	10 D	10 L	12 D	12 L	South Lat. Wall Facing
NE	16	11	18	12	20	13	22	14	23	15	24	16	24	16	23	16	22	16	SE
E	19	13	21	14	25	16	29	17	30	18	30	19	29	20	28	19	26	18	E
SE	19	13	19	13	22	14	26	16	28	18	29	18	29	17	28	18	26	18	NE
S	16	13	15	11	18	11	18	15	21	14	24	16	25	18	25	18	23	16	N
SW	20	14	19	13	18	12	19	13	22	14	27	17	31	20	32	20	30	20	NW
W	22	14	20	13	19	13	20	13	21	14	26	17	31	20	33	21	32	21	W
NW	18	12	16	11	17	11	17	11	18	12	21	14	25	17	27	18	26	18	SW
N	13	10	12	9	13	9	13	10	15	11	16	12	18	13	18	14	18	14	S

Group J[a]

NE	18	13	17	12	18	13	19	15	21	13	22	14	23	15	23	16	16	0.10
E	22	16	20	14	21	15	24	14	26	16	28	17	29	17	28	19	19	
SE	21	17	20	13	20	14	21	15	24	15	26	16	28	17	28	19	18	
S	19	15	17	11	16	11	16	12	17	12	20	13	22	15	24	16	16	
SW	24	16	22	15	20	13	19	13	19	13	21	14	23	15	23	18	19	9
W	26	17	24	16	22	14	20	13	20	13	21	14	29	16	28	18	20	
NW	21	15	19	13	18	12	17	11	17	11	17	12	24	13	22	15	17	
N	15	11	14	11	13	10	13	10	13	10	14	10	15	11	17	12	13	

Group K[a]

NE	19	14	19	13	19	13	20	13	21	14	21	14	22	15	22	15	18	0.08
E	23	16	22	15	23	16	24	15	26	16	25	17	27	18	28	18	19	
SE	23	15	22	15	22	14	22	15	24	15	25	16	26	17	22	16	17	
S	20	14	19	13	18	12	18	12	18	13	20	13	21	15	17	12	15	
SW	25	16	23	15	22	14	21	14	22	14	23	15	24	16	27	18	18	11
W	26	17	24	16	23	13	22	15	23	14	24	16	27	17	28	18	19	
NW	21	15	20	14	19	13	18	13	19	12	20	15	22	15	23	16	17	
N	15	11	14	11	14	10	14	10	14	11	15	12	16	12	16	12	13	

Group L[a]

NE	18	13	18	13	19	13	20	13	21	14	22	15	23	15	22	16	15	0.08
E	22	15	22	14	23	14	21	14	27	17	28	18	28	18	27	18	18	
SE	25	16	23	15	22	12	22	18	25	16	27	17	27	16	24	16	18	
S	20	14	19	13	18	13	18	13	19	13	23	15	23	13	23	16	16	
SW	23	15	21	14	20	13	20	13	21	13	23	17	26	17	28	18	18	8
W	25	16	23	15	21	14	21	14	22	14	24	15	26	15	30	19	19	
NW	20	14	19	13	18	12	18	13	18	12	19	13	21	12	24	16	16	
N	14	13	14	12	13	10	13	10	14	10	15	12	16	12	17	13	13	

Table 1-20A. Total Equivalent Temperature Differentials for Calculating Heat Gain through Sunlit Walls (Continued)

Group M[a]

North Latitude Wall Facing	A.M. 8 D	A.M. 8 L	A.M. 10 D	A.M. 10 L	12 D	12 L	P.M. 2 D	P.M. 2 L	4 D	4 L	6 D	6 L	8 D	8 L	10 D	10 L	12 D	12 L	Amplitude Decrement Factor, λ	Time Lag, δ hr	South Latitude Wall Facing
NE	20	14	20	14	19	13	20	14	20	14	20	14	21	14	21	14	22	15	0.05	12	SE
E	25	16	24	16	24	16	24	16	25	16	25	16	26	17	27	17	27	17			E
SE	24	16	23	15	23	16	23	15	25	15	25	16	25	16	25	16	26	17			NE
S	21	14	20	14	19	13	19	13	19	13	19	13	20	14	21	14	21	15			N
SW	25	17	25	16	24	16	23	15	22	15	22	15	23	15	24	16	25	17			NW
W	27	17	26	17	25	16	24	15	23	13	23	15	24	13	25	16	26	17			W
NW	22	15	21	14	20	14	20	13	20	13	20	13	19	13	20	14	21	15			SW
N	15	12	15	12	14	11	14	11	14	11	14	11	15	11	15	11	16	12			S

Exterior color of wall—D = dark, L = light

[a] See Table 1-20B for details of each wall grouping.

Explanation:

$$\left\{ \begin{array}{l}\text{Total heat transmission from solar radiation and} \\ \text{temperature difference between outside and} \\ \text{room air, Btu/(hr)(ft}^2\text{ wall area)} \end{array} \right\} = \left\{ \begin{array}{l}\text{Equivalent temperature} \\ \text{differential from above} \\ \text{table} \end{array} \right\} \times \left\{ \begin{array}{l}\text{Heat transmission coefficient} \\ \text{for wall, Btu/(hr)(ft}^2\text{)(°F)} \end{array} \right\}$$

1. *Application.* These values may be used for all normal air-conditioning estimates; usually without correction (except as noted below) when the load is calculated for the hottest weather.

2. *Corrections.* The values in the table were calculated for an inside temperature of 75°F and an outdoor maximum temperature of 95°F with an outdoor daily range of 21°F. The table remains approximately correct for other outdoor maximums (93 to 102°F) and other outdoor daily ranges (16 to 34°F) provided the outdoor daily average temperature remains approximately 85°F. If the room temperature is different from 75°F and/or the outdoor daily average temperature is different from 85°F, Equation 43 can be used for computing new values or the following rules can be applied:

  a. For room air temperature less than 75°F, add the difference between 75°F and room air temperature; if greater than 75°F, subtract the difference.

  b. For outdoor daily average temperature less than 85°F, subtract the difference between 85°F and the daily average temperature; if greater than 85°F, add the difference. The table values will be approximately correct for the east or west wall in any latitude (0 to 50° North or South) during the hottest weather. Equation 43 should be used for obtaining values for the north or south wall in latitudes other than 40°.

3. *Color of exterior surface of wall.* Use temperature differentials for light walls only when the permanence of the light wall is established by experience. For cream color use the values for light walls. For medium colors interpolate halfway between the dark and light values. Medium colors are medium blue, medium green, bright red, light brown, unpainted wood, natural color concrete, etc. Dark blue, red, brown, green, etc., are considered dark colors.

Reprinted by permission from "ASHRAE Handbook of Fundamentals," ASHRAE, New York, 1972.

Table 1-20B. Description of Wall Constructions

Group	Components	Wt, lb per sq ft	U Value
A	1″ stucco +4″ l.w. concrete block +air space	28.6	0.267
	1″ stucco +air space +2″ insulation	16.3	0.106
B	1″ stucco +4″ common brick	55.9	0.393
	1″ stucco +4″ h.w. concrete	62.5	0.481
C	4″ face brick +4″ l.w. concrete block +1″ insulation	62.5	0.158
	1″ stucco +4″ h.w. concrete +2″ insulation	62.9	0.114
D	1″ stucco +8″ l.w. concrete block +1″ insulation	41.4	0.141
	1″ stucco +2″ insulation +4″ h.w. concrete block	36.6	0.111
E	4″ face brick +4″ l.w. concrete block	62.2	0.333
	1″ stucco +8″ h.w. concrete block	56.6	0.349
F	4″ face brick +4″ common brick	80.5	0.360
	4″ face brick +2″ insulation +4″ l.w. concrete block	62.5	0.103
G	1″ stucco +8″ clay tile +1″ insulation	62.8	0.141
	1″ stucco +2″ insulation +4″ common brick	56.2	0.108
H	4″ face brick +8″ clay tile +1″ insulation	96.4	0.137
	4″ face brick +8″ common brick	129.6	0.280
	1″ stucco +12″ h.w. concrete	155.9	0.365
	4″ face brick +2″ insulation +4″ common brick	89.8	0.106
	4″ face brick +2″ insulation +4″ h.w. concrete	96.5	0.111
	4″ face brick +2″ insulation +8″ h.w. concrete block	90.6	0.102
I	1″ stucco +8″ clay tile +air space	62.6	0.209
	4″ face brick +air space +4″ h.w. concrete block	69.9	0.282
J	face brick +8″ common brick +1″ insulation	129.8	0.145
	4″ face brick +2″ insulation +8″ clay tile	96.5	0.094
	1″ stucco +2″ insulation +8″ common brick	96.3	0.100
K	4″ face brick +air space +8″ clay tile	96.2	0.200
	4″ face brick +2″ insulation +8″ common brick	129.9	0.098
	4″ face brick +2″ insulation +8″ h.w. concrete	143.3	0.107
L	4″ face brick +8″ clay tile +air space	96.2	0.200
	4″ face brick +air space +4″ common brick	89.5	0.265
	4″ face brick +air space +4″ h.w. concrete	96.2	0.301
	4″ face brick +air space +8″ h.w. concrete block	90.2	0.246
	1″ stucco +2″ insulation +12″ h.w. concrete	156.3	0.106
M	4″ face brick +air space +8″ common brick	129.6	0.218
	4″ face brick +air space +12″ h.w. concrete	189.5	0.251
	4″ face brick +2″ insulation +12″ h.w. concrete	189.9	0.104

Reprinted by permission from "ASHRAE Handbook of Fundamentals," ASHRAE, New York, 1972.

Table 1-21. Total Equivalent Temperature Differentials for Calculating Heat Gain through Flat Roofs

Description of Roof Construction[a,b]	Wt, lb per sq ft	U value Btu/(hr)(ft²)(F°)	A.M. 8 D	8 L	10 D	10 L	12 D	12 L	P.M. 2 D	2 L	4 D	4 L	6 D	6 L	8 D	8 L	10 D	10 L	12 D	12 L	λ	6
Light Construction Roofs—Exposed to Sun																						
1" insulation + steel siding	7.4	0.213	28	11	65	31	90	48	95	53	78	45	43	27	8	6	1	1	-3	-3	1.0	0
2" insulation + steel siding	7.8	0.125	24	8	61	29	88	39	96	53	81	46	48	30	10	8	2	2	-3	-3	0.99	1
1" insulation + 1" wood[c]	8.4	0.206	12	0	47	21	77	39	92	50	86	48	61	36	25	16	7	5	-0	-1	0.93	2
2" insulation + 1" wood[c]	8.5	0.122	8	-2	41	18	72	36	90	48	88	49	65	38	30	19	9	7	1	0	0.93	2
1" insulation + 2.5" wood[c]	12.7	0.193	2	-2	23	8	48	23	70	36	79	42	71	40	50	29	29	17	15	9	0.73	3
2" insulation + 2.5" wood[c]	13.1	0.117	1	-2	19	6	43	20	65	33	76	41	72	40	53	31	33	20	18	11	0.68	4
Medium Construction Roofs—Exposed to Sun																						
1" insulation + 4" wood[c]	17.3	0.183	5	0	14	5	31	14	49	24	62	32	65	35	56	31	41	24	29	17	0.51	5
2" insulation + 4" wood[c]	17.8	0.113	4	-1	13	3	31	12	45	23	58	30	63	34	56	30	43	25	32	18	0.48	5
1" insulation + 2" h.w. concrete	28.3	0.206	6	-1	27	11	54	26	74	39	81	44	70	44	45	27	24	15	12	7	0.75	3
2" insulation + 2" h.w. concrete	28.8	0.122	2	-3	23	9	49	23	70	36	79	43	71	40	49	27	28	17	15	9	0.73	3
4" l.w. concrete	17.8	0.213	-2	-4	28	11	59	28	82	43	88	43	74	40	44	27	19	12	6	4	0.87	3
6" l.w. concrete	24.5	0.157	-2	-2	9	2	31	13	55	27	72	38	76	41	64	36	42	25	25	15	0.67	5
8" l.w. concrete	31.2	0.125	6	3	6	1	16	6	32	14	49	24	61	32	63	34	55	31	41	24	0.50	6
Heavy Construction Roofs—Exposed to Sun																						
1" insulation + 4" h.w. concrete	51.6	0.199	7	1	17	6	33	15	50	25	61	32	63	34	53	30	40	23	28	16	0.48	5
2" insulation + 4" h.w. concrete	52.1	0.120	7	1	15	6	30	13	46	23	58	30	61	33	54	30	41	23	31	17	0.45	5
1" insulation + 6" h.w. concrete	75.0	0.193	13	6	17	7	26	12	48	18	48	25	53	28	51	27	43	24	35	19	0.33	6
2" insulation + 6" h.w. concrete	75.4	0.117	15	7	17	7	25	11	36	17	46	23	51	27	50	27	43	24	36	20	0.30	6

Roofs Covered with Water—Exposed to Sun

Outside Air Dew Point (F)	Water Layer Thickness (in.)

Construction												
Light Construction	60	6 1 0	-12 -12	-6 -6	-1 4 7	6 15 17	13 21 23	17 22 22	17 17 16	13 8 5	7 0 -3
	70	6 1 0	-1 -5 -5	0 0 2	4 10 12	11 19 21	18 25 26	21 26 26	21 21 19	17 12 9	12 5 2
Heavy Construction	60	6 1 0	-3 -8 -9	-4 -6 -5	-1 1 2	4 8 10	9 15 16	13 18 19	15 17 16	13 11 10	10 6 4
	70	6 1 0	-2 -2 -2	2 0 1	4 6 8	9 14 16	14 20 21	18 23 23	20 21 21	18 16 15	15 11 9

Reprinted by permission from "ASHRAE Handbook of Fundamentals," ASHRAE, New York, 1972.

a Includes outside surface resistance, ½-in. slag, membrane and ⅜-in. felt on the top and inside surface resistance on the bottom.
b Dark roof, $\alpha/h_o = 0.30$; light roof, $\alpha/h_o = 0.15$.
c Nominal thickness of wood.

Explanation:
$$\left\{ \begin{array}{l} \text{Total heat transmission from solar radiation and} \\ \text{temperature difference between outdoor and} \\ \text{room air, Btu/(hr)(ft}^2\text{) of roof area} \end{array} \right\} = \left\{ \begin{array}{l} \text{Equivalent temperature} \\ \text{differential from above} \\ \text{table} \\ (°F) \end{array} \right\} \times \left\{ \begin{array}{l} \text{Heat transmission coefficient} \\ \text{for summer, Btu/(hr)/(ft}^2\text{)} \\ (°F) \end{array} \right\}$$

1. *Application.* These values may be used for all normal air conditioning estimated; usually without correction (except as noted below) in latitude 0 to 50° North or South when the load is calculated for the hottest weather.

2. *Corrections.* The values in the table were calculated for an inside temperature of 75°F and an outdoor maximum temperature of 95°F with an outdoor daily range of 21°F. The table remains approximately correct for other outdoor maximums (93 to 102°F) and other outdoor daily ranges (16 to 34°F) provided the outdoor daily average temperature remains approximately 85°F. If the room air temperature is different from 75°F and/or the outdoor daily average temperature is different from 85°F, Equation 43 can be used for computing new values or the following rules can be applied:

a. For room air temperature less than 75°F, add the difference between 75°F and room air temperature; if greater than 75°F, subtract the difference.
b. For outdoor daily average temperature less than 85°F, subtract the difference between 85°F and the daily average temperature; if greater than 85°F, add the difference.

3. *Attics or other spaces between the roof and ceiling.* If the ceiling is insulated and a fan is used for positive ventilation in the space between the ceiling and roof, the total temperature differential for calculating the room load may be decreased by 25 per cent.
If the attic space contains a return duct or other air plenum, care should be taken in determining the portion of the heat gain that reaches the ceiling.

4. *Light colors.* Credit should not be taken for light colored roofs except where the permanence of light colors is established by experience, as in rural areas or where there is little smoke.

5. *For solar transmission in other months.* The table values of temperature differentials that were calculated for July 21 will be approximately correct for a roof in the following months:

North Latitude

Latitude (deg)	Months
0	All Months
10	All Months
20	All Months except Nov., Dec., Jan.
30	Mar., Apr., May, June, July, Aug., Sept.
40	April, May, June, July, Aug.
50	May, June, July

South Latitude

Latitude (deg)	Months
0	All Months
10	All Months
20	All Months except May, June, July
30	Sept., Oct., Nov., Dec., Jan., Feb., March
40	Oct., Nov., Dec., Jan., Feb.
50	Nov., Dec., Jan.

Table 1-22A. Coefficients of Transmission U of Flat Masonry Roofs with Built-up Roofing, with and without Suspended Ceilings* (Winter Conditions, Upward Flow)

These Coefficients are expressed in Btu per (hour) (square foot) (Fahrenheit degree difference in temperature between the air on the two sides), and are based upon an outside wind velocity of 15 mph

Example	Example of Substitution
	For addition of roof insulation, see Table 1-23B

Construction (heat flow up)   Resistance (R)

1. Outside surface (15 mph wind)	0.17
2. Built-up roofing—$\frac{3}{8}$ in.	0.33
3. Roof insulation (none)	—
4. Concrete slab (lt. wt. agg.) (2 in.)	2.22
5. Corrugated metal	0
6. Air space	0.85
7. Metal lath and $\frac{3}{4}$ in. plas. (lt. wt. agg.)	0.47
8. Inside surface (still air)	0.61
Total resistance	4.65
$U = 1/R = 1/4.65 =$	0.22

* To adjust U values for the effect of added insulation between framing members see Table 1-23.
Reprinted from "ASHRAE Handbook of Fundamentals," ASHRAE, New York, 1972.

Table 1-22B. Coefficients of Transmission U of Wood Construction Flat Roofs and Ceilings (Winter Conditions, Upward Flow)

Coefficients are expressed in Btu per (hour) (square foot) (Fahrenheit degree difference in temperature between the air on the two sides), and are based upon an outside wind velocity of 15 mph

Example	Resistance (R)	Example of Substitution	
Construction (Heat flow up)		Delete item 3 and furnish R-19 insulation batt in lieu of item 5 air space.	
1. Outside surface (15 mph wind)	0.17	Total resistance	5.83
2. Built-up roofing (⅜ in.)	0.33	Deduct: 3. Roof insulation	1.39
3. Roof insulation ($C = 0.72$)	1.39	5. Air space	0.85
4. Plywood deck (⅝ in.)	0.78		2.24
5. Air space	0.85	Difference	3.59
6. Gypsum wallboard (½ in.)	0.45	Add: 5. R-19 insulation batt	19.00
7. Acoustical tile (½ in.)—glued	1.25		
8. Inside surface (still air)	0.61	Total resistance	22.59
		$U_i = 1/R = 1/22.59 =$	0.04
Total resistance	5.83		
$U = 1/R = 1/5.83 =$	0.17		

Note: Adjustment for heat flow through framing members depends upon dimensions and spacing of framing.

1–41

Reprinted from "ASHRAE Handbook of Fundamentals," ASHRAE, New York, 1972.

Table 1-22C. Coefficients of Transmission U of Metal Construction Flat Roofs and Ceilings (Winter Conditions, Upward Flow)

Coefficients are expressed in Btu per (hour) (square foot) (Fahrenheit degree difference in temperature between the air on the two sides), and are based on upon outside wind velocity of 15 mph

Example		Example of Substitution		
Construction	Resistance (R)			
1. Outside surface (15 mph wind)........	0.17	Replace item 3 with roof insulation (C = 0.36) and items 6 and 7 with metal lath and ¾ in. plas. (lt. wt. agg.)		
2. Built-up roofing (⅜ in.).........	0.33			
3. Roof insulation (C = 0.24).........	4.17	Total resistance........		6.71
4. Metal deck.........	0.00	Deduct 3. Roof insulation (C = 0.24)....	4.17	
5. Air space.*†.........	0.99	6. Metal lath and		
6. Metal lath and		7. ¾ in. plas. (sand agg.)}	0.13	4.30
7. ¾ in. plas. (sand agg.)}	0.13	Difference.........		2.41
8. Inside surface (still air)........	0.92	Add 3. Roof insulation (C = 0.36)....	2.78	
Total resistance........	6.71	6. Metal lath and		
U = 1/R = 1/6.71..........	0.15	7. ¾ in. plas. (lt. wt. agg.)}	0.47	3.25
Note: Adjustment for heat flow through metal framing members depends upon dimensions and spacing.		Total resistance........		5.66
		U = 1/R = 1/5.66		0.18

* If a vapor barrier is used beneath roof insulation it will have a negligible effect on the U value.
† To adjust U values for the effect of added insulation between framing members, see Table 1-23B.
Reprinted by permission from "ASHRAE Handbook of Fundamentals", ASHRAE, New York, 1972.

Table 1-22D. Coefficients of Transmission U of Pitched Roofs*

Coefficients are expressed in Btu per (hour) (square foot) (Fahrenheit degree difference in temperature between the air on the two sides), and are based on an outside wind velocity of 15 mph for heat flow upward and 7.5 mph for heat flow downward

Example	Resistance (R)	Example of Substitution		
Construction (Heat flow up)		Find U value for same construction with heat flow down (summer conditions)		
1. Outside surface (15 mph wind).	0.17			
2. Asphalt shingle roofing.	0.44	Total resistance.		4.58
3. Building paper.	0.06	Deduct 5. Air space (3.5 in. reflective)	2.06	
4. Plywood deck ($\frac{5}{8}$ in.).	0.78	8. Inside surface (still air).	0.62	2.68
5. Air space †(3.5 in., reflective surface).	2.06			
6. Gypsum wallboard ($\frac{1}{4}$ in.).	0.45	Difference.		1.90
7. Inside surface (still air).	0.62	Add 5. Air space (3.5 in., reflective)	8.08	8.84
		8. Inside surface (still air).	0.76	
Total resistance.	4.58	Total resistance.		10.74
$U = 1/R = 1/4.58$.	0.22	$U = 1/R = 1/10.74$.		0.09
Note: Correction for heat flow through framing members depends upon dimensions and spacing.		Note: Correction for heat flow through framing members depends upon dimensions and spacing.		

1–43

Table 1-22D. Coefficients of Transmission U of Pitched Roofs* (Continued)

Example	Resistance (R)	Example of Substitution	
Construction (Heat flow up)		Find U-value for same construction for summer conditions (heat flow down).	
1. Outside surface (15 mph wind)	0.17	Total resistance	3.42
2. Asphalt shingle roofing	0.44	Deduct: 5. Air space	0.90
3. Building paper	0.06	8. Inside surface (still air)	0.62 1.52
4. $\frac{5}{8}$ in. plywood deck	0.78		
5. Air space	0.90	Difference	1.90
6. $\frac{1}{2}$ in. gypsum wallboard	0.45	Add: 5. Air space	0.89
7. Inside surface (still air)	0.62	8. Inside surface (still air)	0.76 1.65
Total resistance	3.42	Total resistance	3.55
$U_i = 1/R = 1/3.42 =$	0.29	$U_i = 1/R = 1/3.55 =$	0.28
Note: Correction for heat flow through framing members depends upon dimensions and spacing.			

* Pitch of roof = 45°.
† To adjust U values for the effect of added insulation between framing members, see Table 1-23B.
Reprinted by permission from "ASHRAE Handbook of Fundamentals," ASHRAE, New York, 1972.

Table 1-23A. Determination of U Values Resulting from
Addition of Insulation to Any Given Building Section

Given Building Section Property[a,b]		Added R[c,d]						
		$R = 4$	$R = 6$	$R = 8$	$R = 12$	$R = 16$	$R = 20$	$R = 24$
U	R	U	U	U	U	U	U	U
1.00	1.00	0.20	0.14	0.11	0.08	0.06	0.05	0.04
0.90	1.11	0.20	0.14	0.11	0.08	0.06	0.05	0.04
0.80	1.25	0.19	0.14	0.11	0.08	0.06	0.05	0.04
0.70	1.43	0.19	0.13	0.11	0.07	0.06	0.05	0.04
0.60	1.67	0.19	0.13	0.10	0.07	0.06	0.05	0.04
0.50	2.00	0.18	0.13	0.10	0.07	0.06	0.05	0.04
0.40	2.50	0.16	0.12	0.10	0.07	0.05	0.05	0.04
0.30	3.33	0.14	0.11	0.09	0.07	0.05	0.04	0.04
0.20	5.00	0.11	0.09	0.08	0.06	0.05	0.04	0.03
0.10	10.00	0.06	0.06	0.06	0.05	0.04	0.04	0.03
0.08	12.50	0.06	0.06	0.05	0.04	0.04	0.03	0.03

[a] For U or R-values not shown in the table, interpolate as necessary.
[b] Enter column 1 with U or R of the design building section.
[c] Under appropriate column heading for Added R, find U-value of resulting design section.
[d] If the insulation occupies a previously considered air space, an adjustment must be made in the given building section R-value.

Reprinted by permission from "ASHRAE Handbook of Fundamentals," ASHRAE, New York, 1972.

Table 1-23B. Determination of U Value Resulting from
Addition of Insulation to Uninsulated Building Sections

U Value of Roof without Roof-Deck Insulation[a]	Conductance C of Roof-Deck Insulation					
	0.12	0.15	0.19	0.24	0.36	0.72
	U	U	U	U	U	U
0.10	0.05	0.06	0.07	0.07	0.08	0.09
0.15	0.07	0.08	0.08	0.09	0.11	0.12
0.20	0.08	0.09	0.10	0.11	0.13	0.16
0.25	0.08	0.09	0.11	0.12	0.15	0.19
0.30	0.09	0.10	0.12	0.13	0.16	0.21
0.35	0.09	0.10	0.12	0.14	0.18	0.24
0.40	0.09	0.11	0.13	0.15	0.19	0.26
0.50	0.10	0.12	0.14	0.16	0.21	0.29
0.60	0.10	0.12	0.14	0.17	0.22	0.33
0.70	0.10	0.12	0.15	0.18	0.24	0.35

[a] Interpolation or mild extrapolation may be used.

Reprinted by permission from "ASHRAE Handbook of Fundamentals," ASHRAE, New York, 1972.

Table 1-24. Solar Position and Intensity; Solar Heat Gain Factor* for 24° North Latitude

Date	Solar Time A.M.	Alt.	Azimuth	Direct Normal Irradiation, Btuh/sq ft	N	NE	E	SE	S	SW	W	NW	Hor.	Solar Time P.M.
Jan 21	7	4.8	65.6	70	2	20	61	63	25	2	2	2	4	5
	8	16.9	58.3	239	11	41	190	218	114	11	11	11	55	4
	9	27.9	48.8	287	18	22	190	253	166	19	18	18	120	3
	10	37.2	36.1	308	23	23	144	245	200	37	23	23	172	2
	11	43.6	19.6	317	26	26	72	211	220	94	26	26	204	1
	12	46.0	0.0	320	27	27	28	160	227	160	28	27	215	12
				Half Day Totals	93	142	662	1064	833	160	93	93	660	
Feb 21	7	9.0	73.9	153	6	67	141	128	33	6	6	6	16	5
	8	21.9	66.4	261	14	80	220	224	89	14	14	14	83	4
	9	33.9	56.8	297	21	45	208	243	133	22	21	21	153	3
	10	44.5	43.5	313	26	27	157	229	165	28	26	26	205	2
	11	52.2	24.5	321	29	29	80	191	185	67	29	29	238	1
	12	55.2	0.0	323	30	30	31	133	192	133	31	30	249	12
				Half Day Totals	111	269	833	1095	701	199	111	111	817	
Mar 21	7	13.7	83.8	194	10	115	186	145	17	9	9	9	36	5
	8	27.2	76.8	267	18	124	234	204	48	18	18	18	111	4
	9	40.2	67.9	295	24	85	215	214	82	24	24	24	180	3
	10	52.3	54.8	308	29	41	162	195	111	30	29	29	232	2
	11	61.9	33.4	315	32	33	84	154	130	42	32	32	264	1
	12	66.0	0.0	317	33	33	35	95	137	95	35	33	275	12
				Half Day Totals	130	431	922	981	457	163	129	129	960	
Apr 21	6	4.7	100.6	40	5	33	39	22	2	2	2	2	3	6
	7	18.3	94.9	203	19	151	198	127	15	14	14	14	58	5
	8	32.0	89.0	257	24	159	229	165	24	22	22	22	132	4
	9	45.6	81.9	281	29	126	209	169	39	28	28	28	196	3
	10	59.0	71.8	293	34	75	158	148	56	33	33	33	245	2
	11	71.1	51.6	298	36	39	85	107	70	38	36	36	275	1
	12	77.6	0.0	300	37	37	35	58	75	58	39	37	284	12
				Half Day Totals	166	611	953	780	245	164	155	154	1053	
May 21	6	8.0	108.4	85	25	79	83	38	5	5	5	5	12	6
	7	21.2	103.2	203	43	171	196	105	17	17	17	17	72	5
	8	34.6	98.5	248	38	178	218	132	26	24	24	24	142	4
	9	48.3	93.6	269	35	150	198	132	33	31	31	31	201	3
	10	62.0	87.7	280	37	102	150	111	38	35	35	35	247	2
	11	75.5	76.9	286	40	55	83	74	44	39	38	38	274	1
	12	86.0	0.0	287	41	41	41	44	46	44	41	41	282	12
				Half Day Totals	237	754	950	616	186	173	171	171	1090	
June 21	6	9.3	111.6	97	35	93	94	38	7	7	7	7	17	6
	7	22.3	106.8	200	55	176	192	94	18	18	18	18	77	5
	8	35.5	102.6	242	49	184	212	117	26	26	26	26	144	4
	9	49.0	98.7	262	43	158	192	116	33	31	31	31	201	3
	10	62.6	95.0	273	41	113	146	95	38	36	36	36	245	2
	11	76.3	90.7	279	41	64	82	63	41	39	38	39	271	1

Solar heat gain through sheet glass — hourly values (Btu/hr·ft²). Columns read left to right for the A.M. orientations N, NE, E, SE, S, SW, W, NW and HOR; the same rows read as P.M. give the mirror orientations. Times are A.M. (left) and P.M. (right).

Date	A.M.	Solar Alt	Solar Azm	Half Day Totals	N	NE	E	SE	S	SW	W	NW	HOR	P.M.
(June 21)	12	89.5	0.0	280	42	42	42	42	43	42	42	42	279	12
	Half Day Totals				282	804	936	544	184	177	176	177	1095	
July 21	6	8.2	109.2	81	25	76	80	36	17	17	5	5	13	6
	7	21.4	103.8	195	45	168	190	101	27	25	17	17	72	5
	8	34.8	99.2	239	40	176	214	128	34	31	25	25	141	4
	9	48.4	94.5	261	37	150	195	128	39	36	31	31	199	3
	10	62.1	89.0	272	39	104	149	108	44	40	36	35	243	2
	11	75.7	79.2	278	41	57	83	73	46	44	40	38	270	1
	12	86.6	0.0	279	42	42	42	44	46	44	42	42	278	12
	Half Day Totals				247	751	933	598	189	176	172	172	1078	
Aug 21	6	5.0	101.3	34	5	34	34	29	2	2	2	5	4	6
	7	18.5	95.6	186	21	144	186	144	16	15	12	12	58	5
	8	32.2	89.7	240	25	155	220	155	26	23	18	18	129	4
	9	45.9	82.9	265	31	126	202	162	39	30	23	23	191	3
	10	59.3	73.0	277	35	77	154	143	55	34	26	26	238	2
	11	71.6	53.2	283	38	42	85	103	67	39	29	27	267	1
	12	78.3	0.0	285	38	39	41	58	72	58	41	39	276	12
	Half Day Totals				176	603	916	742	242	170	162	161	1027	
Sept 21	7	13.7	83.8	172	11	105	169	132	17	11	11	6	35	5
	8	27.2	76.8	248	19	119	222	194	48	19	19	15	108	4
	9	40.2	67.9	277	26	84	207	206	81	26	26	22	174	3
	10	52.3	54.8	292	30	42	158	190	110	32	30	30	224	2
	11	61.9	33.4	298	33	35	84	151	128	44	33	33	256	1
	12	66.0	0.0	301	34	34	37	94	134	94	37	34	267	12
	Half Day Totals				137	417	879	939	451	172	137	136	930	
Oct 21	7	9.1	74.1	137	6	62	129	117	30	6	6	6	16	5
	8	22.0	66.7	246	15	78	211	213	85	15	15	15	82	4
	9	34.1	57.1	284	22	46	202	235	128	22	22	22	150	3
	10	44.7	43.8	300	27	28	153	222	160	29	27	27	201	2
	11	52.5	24.7	308	30	30	79	186	180	66	30	30	233	1
	12	55.5	0.0	311	31	31	32	130	186	130	32	31	244	12
	Half Day Totals				115	265	800	1050	676	198	116	115	802	
Nov 21	7	4.9	65.8	66	2	20	58	60	24	2	2	2	4	5
	8	17.0	58.4	232	12	41	186	213	111	12	12	12	54	4
	9	28.0	48.9	281	18	23	187	249	162	19	18	18	120	3
	10	37.3	36.3	302	23	24	142	241	197	37	23	23	171	2
	11	43.8	19.7	311	26	26	72	209	217	93	26	26	202	1
	12	46.2	0.0	314	27	27	29	158	224	158	29	27	213	12
	Half Day Totals				93	144	651	1046	817	237	94	93	655	
Dec 21	7	3.2	62.6	29	0	7	41	27	12	0	0	0	1	5
	8	14.9	55.3	225	10	18	174	209	118	10	10	10	44	4
	9	25.5	46.0	281	17	22	180	252	174	18	17	17	106	3
	10	34.3	33.7	304	21	24	137	247	209	44	21	21	157	2
	11	40.4	18.2	314	24	25	69	216	230	104	24	24	188	1
	12	42.6	0.0	317	25	27	27	167	237	167	27	25	199	12
	Half Day Totals				83	107	581	1019	851	254	84	83	593	
					N	NE	E	SE	S	SW	W	NW	HOR	→P.M.

Reprinted by permission from "ASHRAE Handbook of Fundamentals," ASHRAE, New York, 1972.
* Total solar heat gains for DS (⅛ in.) sheet glass. Based on a ground reflectance of 0.20.

1–47

Table 1-25. Solar Position and Intensity; Solar Heat Gain Factors* for 32° North Latitude

Date	Solar Time A.M.	Solar Position Alt.	Solar Position Azimuth	Direct Normal Irradiation, Btuh/sq ft	N	NE	E	SE	S	SW	W	NW	Hor.	Solar Time P.M.
Jan 21	7	1.4	65.2	1	0	0	1	1	1	0	0	0	0	5
	8	12.5	56.5	202	8	29	160	189	103	9	8	8	32	4
	9	22.5	46.0	269	15	16	175	246	169	16	15	15	88	3
	10	30.6	33.1	295	19	20	135	249	212	45	19	19	136	2
	11	36.1	17.5	306	22	22	67	221	238	110	22	22	166	1
	12	38.0	0.0	309	23	23	25	174	246	174	25	23	176	12
	Half Day Totals				75	91	529	974	834	262	75	75	509	
Feb 21	7	6.7	72.8	111	4	47	102	95	26	4	4	4	9	5
	8	18.5	63.8	244	12	64	205	217	95	12	12	12	63	4
	9	29.3	52.8	287	19	32	199	248	149	19	19	19	127	3
	10	38.5	38.9	305	23	24	151	241	189	31	23	23	176	2
	11	44.9	21.0	314	26	26	76	208	213	87	26	26	207	1
	12	47.2	0.0	316	27	27	29	174	221	155	29	27	217	12
	Half Day Totals				97	207	749	1091	780	227	98	97	689	
Mar 21	7	12.7	81.9	184	9	105	176	142	19	9	9	9	31	5
	8	25.1	73.0	260	17	107	227	209	62	17	17	17	99	4
	9	36.8	62.1	289	23	64	210	227	107	23	23	23	163	3
	10	47.5	47.5	304	27	30	158	215	144	29	27	27	211	2
	11	55.0	26.8	310	30	31	82	179	168	58	30	30	242	1
	12	58.0	0.0	312	31	31	33	122	176	122	33	31	252	12
	Half Day Totals				122	368	891	1054	588	191	123	122	872	
Apr 21	6	6.1	99.9	66	3	54	65	37	3	3	3	3	7	6
	7	18.8	92.2	206	17	147	201	136	15	14	14	14	61	5
	8	31.5	84.0	256	23	144	228	178	30	22	22	22	130	4
	9	43.9	74.2	278	28	103	206	188	58	27	27	27	189	3
	10	55.7	60.3	290	32	52	156	173	87	33	32	32	234	2
	11	65.4	37.5	296	34	36	83	135	108	40	34	34	263	1
	12	69.6	0.0	298	35	35	38	82	115	82	38	35	272	12
	Half Day Totals				159	559	965	898	359	174	150	149	1022	
May 21	6	10.4	107.2	118	32	108	116	55	8	8	8	8	21	6
	7	22.8	100.1	211	35	170	204	118	18	18	18	18	81	5
	8	35.4	92.9	249	29	165	220	149	27	25	25	25	146	4
	9	48.1	84.7	269	32	128	198	155	37	30	30	30	201	3
	10	60.6	73.3	279	36	76	150	138	54	35	35	35	243	2
	11	72.0	51.9	285	38	41	82	102	68	39	37	37	269	1
	12	78.0	0.0	286	38	39	41	59	74	59	41	39	277	12
	Half Day Totals				217	697	983	747	248	181	172	171	1100	
June 21	6	12.2	110.2	130	44	123	127	55	10	10	10	10	28	6
	7	24.3	103.4	209	46	176	201	108	19	19	19	19	88	5
	8	36.9	96.8	244	36	171	214	135	28	26	26	26	151	4
	9	49.6	89.4	263	34	136	193	139	35	32	32	32	203	3
	10	62.2	79.7	273	38	86	146	122	45	36	36	36	244	2
	11	74.2	60.9	278	40	46	81	88	56	40	38	38	268	1

	12	81.5	0.0	280	40	41	42	52	60	52	42	41	276	12
			Half Day Totals		252	744	972	672	222	186	180	180	1119	
July 21	6	10.7	107.7	113	33	105	112	53	8	8	8	8	22	6
	7	23.1	100.6	203	37	167	198	114	19	18	18	18	81	5
	8	35.7	93.6	241	31	163	216	145	28	26	26	26	145	4
	9	48.4	85.5	261	34	128	195	150	37	31	31	31	199	3
	10	60.9	74.3	271	37	78	148	134	53	35	35	35	240	2
	11	72.4	53.3	277	39	43	82	99	66	40	38	38	265	1
	12	78.6	0.0	278	40	40	42	58	71	58	42	40	273	12
			Half Day Totals		227	694	964	724	245	184	176	175	1089	
Aug 21	6	6.5	100.5	59	9	50	59	34	3	3	3	3	7	6
	7	19.1	92.8	189	18	140	189	127	16	15	15	15	61	5
	8	31.8	84.7	239	25	141	219	170	30	23	23	23	127	4
	9	44.3	75.0	263	30	104	200	180	56	29	29	29	185	3
	10	56.1	61.3	275	33	55	152	167	84	34	33	33	229	2
	11	66.0	38.4	281	36	38	83	131	104	41	36	36	256	1
	12	70.3	0.0	283	37	37	40	80	111	80	40	37	265	12
			Half Day Totals		169	552	929	858	349	180	159	158	999	
Sep 21	7	12.7	81.9	163	10	95	159	128	19	9	9	9	30	5
	8	25.1	73.0	240	18	103	215	199	60	18	18	18	96	4
	9	36.8	62.1	272	24	64	202	218	105	24	24	24	158	3
	10	47.3	47.5	287	29	32	154	208	141	30	29	29	204	2
	11	55.0	26.8	294	31	32	81	174	164	59	31	31	234	1
	12	58.0	0.0	296	32	32	34	120	171	120	34	32	244	12
			Half Day Totals		128	355	846	1004	575	194	128	127	844	
Oct 21	7	6.8	73.1	98	4	43	92	85	23	4	4	4	9	5
	8	18.7	64.0	229	13	63	195	205	90	13	13	13	62	4
	9	29.5	53.0	273	19	33	193	239	144	20	19	19	125	3
	10	38.7	39.1	292	24	25	147	234	183	32	24	24	173	2
	11	45.1	21.1	301	27	27	75	202	206	85	27	27	203	1
	12	47.5	0.0	304	28	28	30	151	214	151	30	28	213	12
			Half Day Totals		100	205	718	1044	750	225	101	100	677	
Nov 21	7	1.5	65.4	1	0	0	1	1	0	0	0	0	0	5
	8	12.7	56.6	196	9	29	156	183	100	9	9	9	32	4
	9	22.6	46.1	262	15	17	172	241	166	16	15	15	87	3
	10	30.8	33.2	288	20	20	134	244	209	45	20	20	135	2
	11	36.2	17.6	300	22	22	67	218	234	108	22	22	165	1
	12	38.2	0.0	303	23	23	25	171	243	171	25	23	175	12
			Half Day Totals		76	92	521	955	820	258	77	76	505	
Dec 21	8	10.3	53.8	176	7	18	135	166	96	7	7	7	22	4
	9	19.8	43.6	257	13	14	162	238	171	15	13	13	72	3
	10	27.6	31.2	287	18	18	127	246	217	52	18	18	119	2
	11	32.7	16.4	300	20	20	63	222	243	116	20	20	148	1
	12	34.6	0.0	304	21	21	23	177	252	177	23	21	158	12
			Half Day Totals		67	76	482	947	844	273	68	67	440	↑
					N	NW	W	SW	S	SE	E	NE	HOR.	←P.M.

Reprinted by permission from "ASHRAE Handbook of Fundamentals," ASHRAE, New York, 1972.

* Total solar heat gains for DS (1/8 in.) sheet glass. Based on a ground reflectance of 0.20.

Table 1-26. Solar Position and Intensity; Solar Heat Gain Factors* for 40° North Latitude

Date	Solar Time A.M.	Solar Position Alt.	Solar Position Azimuth	Direct Normal Irradiation, Btuh/sq ft	N	NE	E	SE	S	SW	W	NW	Hor.	Solar Time P.M.
Jan 21	8	8.1	55.3	141	5	17	111	133	75	5	5	5	13	4
	9	16.8	44.0	238	11	12	154	224	160	13	11	11	54	3
	10	23.8	30.9	274	16	16	123	241	213	51	16	16	96	2
	11	28.4	16.0	289	18	18	61	222	244	118	18	18	123	1
	12	30.0	0.0	293	19	19	20	179	254	179	20	19	133	12
				Half Day Totals	59	68	449	903	815	271	59	59	353	
Feb 21	7	4.3	72.1	55	1	22	50	47	13	1	1	1	13	5
	8	14.8	61.6	219	10	50	183	199	94	10	10	10	43	4
	9	24.3	49.7	271	16	22	186	245	157	17	16	16	98	3
	10	32.1	35.4	293	20	21	142	247	203	38	20	20	143	2
	11	37.3	18.6	303	23	23	71	219	231	103	23	23	171	1
	12	39.2	0.0	306	24	24	25	170	241	170	25	24	180	12
				Half Day Totals	81	144	634	1035	813	250	81	81	546	
Mar 21	7	11.4	80.2	171	8	93	163	135	21	8	8	8	26	5
	8	22.5	69.6	250	15	91	218	211	73	15	15	15	85	4
	9	32.8	57.3	281	21	46	203	236	128	21	21	21	143	3
	10	41.6	41.9	297	25	26	153	229	171	28	25	25	186	2
	11	47.7	22.6	304	28	28	78	198	197	77	28	28	213	1
	12	50.0	0.0	306	28	28	30	145	206	145	30	28	223	12
				Half Day Totals	112	310	849	1100	692	218	112	112	764	
Apr 21	6	7.4	98.9	89	17	72	88	52	5	4	4	4	11	6
	7	18.9	89.5	207	11	141	201	143	16	14	14	14	61	5
	8	30.3	79.2	253	16	128	225	189	41	21	21	21	124	4
	9	41.3	67.2	275	22	80	203	204	83	26	26	26	177	3
	10	51.2	51.4	286	26	37	153	194	121	32	30	30	218	2
	11	58.7	29.2	292	30	34	81	161	146	52	33	33	244	1
	12	61.6	0.0	294	33	33	36	108	155	108	36	33	253	12
				Half Day Totals	153	509	969	1003	489	196	112	145	962	
May 21	5	1.9	114.7	1	1	1	1	1	0	0	0	0	0	7
	6	12.7	105.6	143	35	128	141	71	10	10	10	10	30	6
	7	24.0	96.6	216	28	165	209	131	20	18	18	18	87	5
	8	35.4	87.2	249	27	149	220	164	29	25	25	25	146	4
	9	46.8	76.0	267	31	105	197	175	53	30	30	30	196	3
	10	57.5	60.9	277	34	54	148	163	83	35	34	34	234	2
	11	66.2	37.1	282	36	38	81	130	105	42	36	36	258	1
	12	70.0	0.0	284	37	37	40	82	112	82	40	37	265	12
				Half Day Totals	203	643	1002	1003	356	194	171	170	1083	
June 21	5	4.2	117.3	21	10	20	20	6	6	2	2	2	6	7
	6	14.8	108.4	154	47	142	151	70	12	12	12	12	39	6
	7	26.0	99.7	215	37	172	207	122	21	20	20	20	97	5
	8	37.4	90.7	246	29	156	215	152	29	26	26	26	153	4
	9	48.8	80.2	262	33	113	192	161	45	31	31	31	201	3
	10	59.8	65.8	272	35	62	145	148	69	36	35	35	237	2

					N	NW	W	SW	S	SE	E	NE	HOR.	
	11	69.2	41.9	276	37	40	80	116	88	41	37	37	260	1
	12	73.5	0.0	279	38	38	41	71	95	71	41	38	267	12
	Half Day Totals				242	714	1019	810	311	197	181	180	1121	
July 21	5	2.3	115.2	2	0	2	1	0	0	0	0	0	0	7
	6	13.1	106.1	137	37	125	137	68	10	10	10	10	31	6
	7	24.3	97.2	208	30	163	204	127	20	19	19	19	88	5
	8	35.8	87.8	241	28	148	216	160	29	26	26	26	145	4
	9	47.2	76.7	259	32	106	194	170	52	31	31	31	194	3
	10	57.9	61.7	269	35	56	146	159	80	36	35	35	231	2
	11	66.7	37.9	274	37	39	81	127	102	42	37	37	255	1
	12	70.6	0.0	276	38	38	41	80	109	80	41	38	262	12
	Half Day Totals				211	645	986	850	347	197	177	176	1074	
Aug 21	6	7.9	99.5	80	12	67	82	48	5	5	5	5	11	6
	7	19.3	90.0	191	17	135	191	135	17	15	15	15	62	5
	8	30.7	79.9	236	23	126	216	180	40	22	22	22	122	4
	9	41.8	67.9	259	28	82	197	196	79	28	28	28	174	3
	10	51.7	52.1	271	32	40	149	187	116	34	32	32	213	2
	11	59.3	29.7	277	34	35	81	156	140	52	34	34	238	1
	12	62.3	0.0	279	35	35	38	105	149	105	38	35	247	12
	Half Day Totals				161	503	936	961	471	202	154	153	945	
Sep 21	7	11.4	80.2	149	8	84	146	121	21	8	8	8	25	5
	8	22.5	69.6	230	16	87	205	199	71	16	16	16	82	4
	9	32.8	57.3	263	22	47	195	226	124	23	22	22	138	3
	10	41.6	41.9	279	26	28	148	221	165	30	26	26	180	2
	11	47.7	22.6	287	29	29	77	192	191	77	29	29	206	1
	12	50.0	0.0	290	30	30	32	141	200	141	32	30	215	12
	Half Day Totals				116	300	803	1045	672	221	117	116	738	
Oct 21	7	4.5	72.3	48	1	20	45	41	12	1	1	1	3	5
	8	15.0	61.9	203	10	49	173	187	88	10	10	10	43	4
	9	24.5	49.8	257	17	23	180	235	151	18	17	17	96	3
	10	32.4	35.6	280	21	22	139	238	196	38	21	21	140	2
	11	37.6	18.7	290	23	23	70	212	224	100	23	23	167	1
	12	39.5	0.0	293	24	24	26	165	234	165	26	24	177	12
	Half Day Totals				83	143	610	989	783	245	84	83	535	
Nov 21	8	8.2	55.4	136	5	17	107	128	72	5	5	5	14	4
	9	17.0	44.1	232	12	13	151	219	156	13	12	12	54	3
	10	24.0	31.0	267	16	16	122	237	209	50	16	16	96	2
	11	28.6	16.1	283	19	19	61	218	240	116	19	19	123	1
	12	30.2	0.0	287	19	19	21	176	250	176	21	19	132	12
	Half Day Totals				61	71	442	884	798	267	62	61	353	
Dec 21	8	5.5	53.0	88	2	7	67	83	49	3	2	2	6	4
	9	14.0	41.9	217	9	10	135	205	151	12	9	9	39	3
	10	20.7	29.4	261	14	14	113	232	210	55	14	14	77	2
	11	25.0	15.2	279	16	16	56	217	242	120	16	16	103	1
	12	26.6	0.0	284	17	17	18	177	253	177	18	17	113	12
	Half Day Totals				49	54	380	831	781	273	50	49	282	
					N	NW	W	SW	S	SE	E	NE	HOR.	←P.M.

Reprinted by permission from "ASHRAE Handbook of Fundamentals," ASHRAE, New York, 1972.
* Total solar heat gains for DS (⅛ in.) sheet glass. Based on a ground reflectance of 0.20.

Table 1-27. Solar Position and Intensity; Solar Heat Gain Factors* for 48° North Latitude

Date	Solar Time A.M.	Alt.	Azimuth	Direct Normal Irradiation, Btuh/sq ft	N	NE	E	SE	S	SW	W	NW	Hor.	Solar Time P.M.
Jan 21	8	3.5	54.6	36	1	4	28	34	19	1	1	1	2	4
	9	11.0	42.6	185	7	8	117	176	128	9	7	7	25	3
	10	16.9	29.4	239	11	11	105	216	195	50	11	11	55	2
	11	20.7	15.1	260	14	14	52	208	233	115	14	14	77	1
	12	22.0	0.0	267	15	15	16	171	245	171	16	15	85	12
	Half Day Totals				41	44	319	735	705	256	41	41	202	
Feb 21	7	1.8	71.7	3	0	3	3	3	0	0	0	0	3	5
	8	10.9	60.0	180	7	36	149	166	82	7	7	7	24	4
	9	19.0	47.3	247	13	15	168	230	155	14	13	13	66	3
	10	25.5	33.0	275	17	17	131	242	207	43	17	17	105	2
	11	29.7	17.0	288	19	19	65	221	240	112	20	19	129	1
	12	31.2	0.0	291	20	20	21	176	251	176	21	20	138	12
	Half Day Totals				65	88	508	936	803	260	65	65	392	
Mar 21	7	10.0	78.7	152	7	80	145	123	22	7	7	7	20	5
	8	19.5	66.8	235	13	75	204	206	81	13	13	13	67	4
	9	28.2	53.4	270	19	33	193	239	142	19	19	19	117	3
	10	35.4	37.8	287	22	24	146	237	189	33	22	22	156	2
	11	40.3	19.8	295	24	24	73	210	218	94	27	24	180	1
	12	42.0	0.0	297	25	25	27	161	228	161	27	25	188	12
	Half Day Totals				98	256	790	1111	765	244	99	98	634	
Apr 21	6	8.6	97.8	108	12	86	107	64	6	6	6	6	14	6
	7	18.6	86.7	205	15	133	200	149	17	14	14	14	60	5
	8	28.5	74.9	247	20	111	219	197	55	20	20	20	114	4
	9	37.8	61.2	269	25	60	198	216	107	25	25	25	161	3
	10	45.8	44.6	281	28	30	148	210	150	30	28	28	197	2
	11	51.5	24.0	287	30	30	78	181	177	69	30	30	219	1
	12	53.6	0.0	289	31	31	33	132	187	132	33	31	227	12
	Half Day Totals				143	459	961	1086	604	225	139	138	879	
May 21	5	5.2	114.3	41	16	39	38	13	2	2	2	2	4	7
	6	14.7	103.7	162	35	140	160	84	12	12	12	12	39	6
	7	24.6	93.0	218	23	158	212	142	21	19	19	19	91	5
	8	34.6	81.6	248	26	132	218	178	38	24	24	24	142	4
	9	44.3	68.3	264	29	82	194	192	77	29	29	29	186	3
	10	53.0	51.3	274	32	39	145	184	116	34	32	32	219	2
	11	59.5	28.6	279	34	35	79	155	142	54	38	34	240	1
	12	62.0	0.0	280	35	35	38	106	150	106	38	35	247	12
	Half Day Totals				209	637	1058	1002	483	220	170	169	1043	
June 21	5	7.9	116.5	77	35	76	72	23	5	5	5	12	7	
	6	17.2	106.2	172	46	154	169	84	14	14	14	14	51	6
	7	27.0	95.8	219	29	165	211	135	22	21	21	21	102	5
	8	37.1	84.6	245	28	139	215	167	34	26	26	26	152	4
	9	46.9	71.6	260	31	90	190	179	66	31	31	31	193	3
	10	55.8	54.8	269	34	45	142	171	101	36	34	34	225	2

Table 1-53 — Total Solar Heat Gains (Btu/hr·ft²), North Latitude 40°

Columns: Solar Time (A.M.) | Solar Altitude | Solar Azimuth | Direct Normal | N | NE | E | SE | S | SW | W | NW | HOR. | Solar Time (P.M.)

June 21 (continued)

A.M.	Alt.	Az.	D.N.	N	NE	E	SE	S	SW	W	NW	HOR.	P.M.
11	62.7	31.2	273	36	37	78	142	125	49	36	36	245	1
12	65.4	0.0	275	36	36	39	96	134	96	39	36	252	12
Half Day Totals				**258**	**728**	**1098**	**952**	**434**	**224**	**186**	**185**	**1105**	

July 21

A.M.	Alt.	Az.	D.N.	N	NE	E	SE	S	SW	W	NW	HOR.	P.M.
5	5.7	114.7	42	18	42	40	14	2	2	2	2	5	7
6	15.2	104.1	155	36	138	156	82	12	12	12	12	41	6
7	25.1	93.5	211	24	156	207	138	21	20	20	20	92	5
8	35.1	82.1	240	27	132	214	174	38	25	25	25	142	4
9	44.8	68.8	256	30	83	191	187	75	30	30	30	184	3
10	53.5	51.9	266	33	41	143	180	113	35	33	33	217	2
11	60.1	29.0	271	35	37	79	151	138	53	35	35	238	1
12	62.6	0.0	272	36	36	39	104	146	104	39	36	245	12
Half Day Totals				**218**	**643**	**1044**	**978**	**472**	**222**	**175**	**174**	**1040**	

Aug 21

A.M.	Alt.	Az.	D.N.	N	NE	E	SE	S	SW	W	NW	HOR.	P.M.
6	9.1	98.3	98	13	81	100	60	7	6	6	6	16	6
7	19.1	87.2	189	16	127	189	140	18	15	15	15	61	5
8	29.0	75.4	231	21	110	211	188	53	21	21	21	113	4
9	38.4	61.8	253	26	62	192	208	102	26	26	26	159	3
10	46.4	45.1	265	30	32	145	202	144	32	32	30	193	2
11	52.2	24.3	271	32	32	78	175	171	68	35	32	215	1
12	54.3	0.0	273	33	33	35	128	180	128	39	33	222	12
Half Day Totals				**152**	**454**	**927**	**1040**	**584**	**227**	**147**	**152**	**868**	

Sep 21

A.M.	Alt.	Az.	D.N.	N	NE	E	SE	S	SW	W	NW	HOR.	P.M.
7	10.0	78.7	131	14	71	128	108	21	7	7	14	19	5
8	19.5	66.8	215	20	72	191	193	77	14	14	20	65	4
9	28.2	53.4	251	23	33	184	227	136	20	20	23	113	3
10	35.4	37.8	269	26	25	141	228	182	34	23	26	151	2
11	40.3	19.8	277	32	26	73	203	211	92	26	32	174	1
12	42.0	0.0	280	33	26	28	156	220	156	28	33	182	12
Half Day Totals				**104**	**246**	**744**	**1050**	**736**	**242**	**104**	**146**	**612**	

Oct 21

A.M.	Alt.	Az.	D.N.	N	NE	E	SE	S	SW	W	NW	HOR.	P.M.
7	2.0	71.9	3	8	1	3	3	0	0	3	1	0	5
8	11.2	60.2	165	13	35	139	155	76	15	16	8	24	4
9	19.3	47.4	232	17	16	161	219	147	43	18	13	66	3
10	25.7	33.1	261	20	18	127	233	199	109	20	17	103	2
11	30.0	17.1	274	20	20	64	213	231	168	20	20	127	1
12	31.5	0.0	278	20	20	22	170	242	170	22	20	136	12
Half Day Totals				**104**	**104**	**488**	**895**	**768**	**316**	**91**	**91**	**387**	

Nov 21

A.M.	Alt.	Az.	D.N.	N	NE	E	SE	S	SW	W	NW	HOR.	P.M.
8	3.6	54.7	36	7	4	28	34	19	9	7	4	2	4
9	11.2	42.7	178	12	8	115	171	125	49	12	8	25	3
10	17.1	29.5	232	14	12	104	211	191	113	14	12	55	2
11	20.9	15.1	254	15	14	52	204	228	168	16	14	77	1
12	22.2	0.0	260	41	45	16	168	240	202	42	41	85	12
Half Day Totals				**41**	**45**	**241**	**719**	**690**	**252**	**42**	**41**	**202**	

Dec 21

A.M.	Alt.	Az.	D.N.	N	NE	E	SE	S	SW	W	NW	HOR.	P.M.
9	8.0	40.9	140	5	5	86	133	100	49	9	5	13	3
10	13.6	28.2	214	9	9	91	194	179	111	12	9	37	2
11	17.3	14.4	242	12	12	46	195	220	163	14	12	57	1
12	18.6	0.0	250	12	14	14	163	233	163	16	12	64	12
Half Day Totals				**32**	**32**	**202**	**621**	**623**	**244**	**33**	**32**	**139**	

→ P.M.

*Total solar heat gains for DS (⅛ in.) sheet glass. Based on a ground reflectance of 0.20.
Reprinted by permission from "ASHRAE Handbook of Fundamentals," ASHRAE, New York, 1972.

1–53

Table 1-28. Solar Position and Intensity; Solar Heat Gain Factors* for 56° North Latitude

Date	Solar Time A.M.	Alt.	Azimuth	Direct Normal Irradiation, Btuh/sq ft	N	NE	E	SE	S	SW	W	NW	Hor.	Solar Time P.M.
Jan 21	9	5.0	41.8	77	2	2	48	74	54	3	2	2	5	3
	10	9.9	28.5	170	6	6	73	156	143	38	6	6	20	2
	11	12.9	14.5	206	9	9	39	169	190	96	9	9	34	1
	12	14.0	0.0	216	9	9	10	143	205	143	10	9	39	12
				Half Day Totals	21	21	168	475	489	205	22	21	78	
Feb 21	8	6.9	59.0	115	4	20	94	107	54	4	4	4	9	4
	9	13.5	45.6	207	13	10	139	197	136	10	9	9	36	3
	10	18.7	31.2	245	15	13	115	223	196	45	13	13	64	2
	11	22.2	15.9	262	15	15	56	210	232	112	15	15	84	1
	12	23.2	0.0	267	15	15	17	171	244	171	17	15	91	12
				Half Day Totals	48	60	405	819	738	252	49	48	239	
Mar 21	7	8.3	77.5	127	5	64	121	104	21	5	5	5	14	5
	8	16.2	64.4	215	11	61	185	194	83	11	11	11	49	4
	9	23.3	50.3	253	16	23	179	233	148	17	16	16	89	3
	10	29.0	34.9	272	19	20	136	238	198	38	19	19	122	2
	11	32.7	17.9	281	21	21	68	215	230	106	21	21	142	1
	12	34.0	0.0	284	22	22	23	170	241	170	23	22	149	12
				Half Day Totals	83	205	712	1080	799	260	83	83	490	
Apr 21	5	1.4	108.8	0	0	0	0	0	0	0	0	0	0	7
	6	9.6	96.5	122	12	96	121	75	7	0	7	7	18	6
	7	18.0	84.1	201	14	123	196	152	21	13	13	13	56	5
	8	26.1	70.9	240	19	95	212	202	68	19	19	19	101	4
	9	33.6	56.3	261	23	44	190	224	126	23	23	23	141	3
	10	39.9	39.7	273	26	27	142	221	172	32	26	26	171	2
	11	44.1	20.7	279	27	27	74	196	201	86	27	27	189	1
	12	45.6	0.0	280	28	28	30	149	211	149	30	28	196	12
				Half Day Totals	132	413	940	1144	699	251	128	128	773	
May 21	4	1.2	125.5	0	0	0	0	0	0	0	0	0	0	8
	5	8.5	113.4	92	36	89	87	33	6	0	6	6	14	7
	6	16.5	101.5	175	32	148	173	97	14	6	13	13	48	6
	7	24.8	89.3	219	21	149	212	152	52	13	19	19	92	5
	8	33.1	76.3	244	24	115	215	189	102	24	24	24	135	4
	9	40.9	61.6	259	27	62	189	206	145	27	27	27	171	3
	10	47.6	44.2	268	30	33	141	200	172	33	30	30	199	2
	11	52.3	23.4	273	32	32	75	174	181	70	32	32	216	1
	12	54.0	0.0	275	33	28	35	129	181	129	35	33	222	12
				Half Day Totals	223	651	1115	1120	602	252	168	167	986	
June 21	4	4.2	127.2	21	13	21	17	40	9	9	9	9	2	8
	5	11.4	115.3	121	52	119	114	97	16	16	16	16	24	7
	6	19.3	103.6	185	42	160	182	147	23	23	21	21	61	6
	7	27.6	91.7	221	24	156	213	181	46	26	26	26	105	5
	8	35.9	78.8	243	27	121	212	195	91	29	29	29	146	4
	9	43.8	64.1	256	29	69	186	189	132	35	32	32	181	3
	10	50.7	46.4	264	32	35	139						208	2

Total solar heat gains for DS (⅛ in.) sheet glass — column reference header

	Solar Alt.	Solar Azm.	Tot.	N	NW	W	SW	S	SE	E	NE	HOR.
	55.6	24.9	268	34	35	76	164	158	64	34	34	225
	57.4	0.0	270	34	34	37	119	167	119	37	34	230
(whole‑day totals)				268	734	1158	1079	560	255	186	185	1067

A.M.	Solar Alt.	Solar Azm.	Tot.	N	NW	W	SW	S	SE	E	NE	HOR.	P.M.
July 21													
4	1.7	125.8	1	—	0	0	0	—	—	0	0	0	8
5	9.0	113.7	91	0	89	87	33	0	0	6	6	15	7
6	17.0	101.9	169	37	145	170	94	14	6	14	14	50	6
7	25.3	89.7	212	34	147	208	148	22	14	20	20	93	5
8	33.6	76.7	236	25	115	211	185	51	25	25	25	135	4
9	41.4	62.0	251	28	63	186	201	99	28	28	28	171	3
10	48.2	44.6	260	31	34	139	196	141	34	31	31	198	2
11	52.9	23.7	265	33	33	76	171	168	70	33	33	215	1
12	54.6	0.0	267	34	34	37	126	177	126	37	34	221	12
Half Day Totals				231	650	1101	1096	589	256	175	174	987	
Aug 21													
5	2.0	109.2	1	—	0	0	0	0	0	0	0	0	7
6	10.2	97.0	112	0	91	114	70	8	7	7	7	19	6
7	18.5	84.5	186	13	119	186	144	21	15	15	15	58	5
8	26.7	71.3	224	16	94	203	192	66	20	20	20	101	4
9	34.3	56.7	245	20	46	184	215	121	25	24	24	139	3
10	40.5	40.0	257	24	29	139	213	165	34	27	27	168	2
11	44.8	20.9	263	27	29	74	189	193	84	29	29	187	1
12	46.3	0.0	265	30	30	32	145	203	145	32	30	193	12
Half Day Totals				142	412	908	1095	673	254	137	136	768	
Sep 21													
7	8.3	77.5	107	5	56	104	90	19	—	5	5	13	5
8	16.2	64.4	194	12	57	171	179	78	12	12	12	48	4
9	23.3	50.3	233	17	24	170	220	140	18	17	17	86	3
10	29.0	34.9	253	20	21	131	227	189	39	20	20	117	2
11	32.7	17.9	263	22	22	67	206	220	102	22	22	137	1
12	34.0	0.0	266	23	23	25	163	231	163	25	23	144	12
Half Day Totals				87	195	664	1013	760	255	88	87	472	
Oct 21													
8	7.1	59.1	103	4	19	86	98	50	—	4	4	10	4
9	13.8	45.7	192	10	11	132	185	128	11	10	10	36	3
10	19.0	31.3	230	13	13	110	212	186	43	13	13	64	2
11	22.3	16.0	247	15	15	55	201	222	108	15	15	83	1
12	23.5	0.0	252	16	16	18	164	234	164	18	16	90	12
Half Day Totals				50	61	386	776	702	244	50	50	238	
Nov 21													
9	5.2	41.9	75	2	2	47	103	53	—	3	2	5	3
10	10.1	28.5	164	7	7	72	152	139	37	7	7	21	2
11	13.1	14.5	200	9	9	39	165	185	93	9	9	34	1
12	14.2	0.0	210	10	10	11	140	200	140	11	10	40	12
Half Day Totals				22	22	166	464	476	199	23	22	79	
Dec 21													
9	1.9	40.5	5	0	0	3	4	3	—	0	0	0	3
10	6.6	27.5	113	3	3	46	103	96	27	3	3	9	2
11	9.5	13.9	165	6	6	29	135	154	78	6	6	19	1
12	10.6	0.0	180	7	7	8	120	171	120	8	7	23	12
Half Day Totals				12	12	76	294	330	162	12	12	39	

↔P.M.

* Total solar heat gains for DS (⅛ in.) sheet glass. Based on a ground reflectance of 0.20.
Reprinted by permission from "ASHRAE Handbook of Fundamentals" ASHRAE, New York 1972.

Table 1-29. Shading Coefficients for Single Glass and Insulating Glass[a]

A. Single Glass

Type of Glass	Nominal Thickness[b]	Solar Trans.[b]	Shading Coefficient	
			$h_0 = 4.0$	$h_0 = 3.0$
Regular Sheet Regular Plate/ Float	$\frac{3}{32}, \frac{1}{8}$ $\frac{1}{4}$ $\frac{3}{8}$ $\frac{1}{2}$	0.87 0.80 0.75 0.71	1.00 0.95 0.91 0.88	1.00 0.97 0.93 0.91
Grey Sheet	$\frac{1}{8}$ $\frac{3}{16}$ $\frac{7}{32}$ $\frac{7}{32}$ $\frac{1}{4}$	0.59 0.74 0.45 0.71 0.67	0.78 0.90 0.66 0.88 0.86	0.80 0.92 0.70 0.90 0.88
Heat-Absorbing Plate/Float[d]	$\frac{3}{16}$ $\frac{1}{4}$ $\frac{3}{8}$ $\frac{1}{2}$	0.52 0.47 0.33 0.24	0.72 0.70 0.56 0.50	0.75 0.74 0.61 0.57

B. Insulating Glass[a]

Type of Glass	Nominal Thickness[c]	Solar Trans.[b]		Shading Coefficient	
		Outer Pane	Inner Pane	$h_0 = 4.0$	$h_0 = 3.0$
Regular Sheet Out, Regular Sheet In	$\frac{3}{32}, \frac{1}{8}$	0.87	0.87	0.90	0.90
Regular Plate/Float Out, Regular Plate/Float In	$\frac{1}{4}$	0.80	0.80	0.83	0.83
Heat-Abs Plate/Float Out, Regular Plate/Float In	$\frac{1}{4}$	0.46	0.80	0.56	0.58

[a] Refers to factory-fabricated units with $\frac{3}{16}$, $\frac{1}{4}$, or $\frac{1}{2}$ in. air space or to prime windows plus storm windows.
[b] Refer to manufacturer's literature for values.
[c] Thickness of each pane of glass, not thickness of assembled unit.
[d] Refers to grey, bronze, and green tinted heat-absorbing plate/float glass.

Table 1-30. Shading Coefficients for Single Glass with Indoor Shading by Venetian Blinds and Roller Shades

Type of Glass	Nominal Thickness[a]	Solar Trans.[b]	Venetian Blinds		Roller Shade		
					Opaque		Translucent
			Medium	Light	Dark	White	Light
Regular Sheet	$\frac{3}{32}$ to $\frac{1}{4}$	0.87–0.80	0.64	0.55	0.59	0.25	0.39
Regular Plate/Float	$\frac{1}{4}$ to $\frac{1}{2}$	0.80–0.71					
Regular Pattern	$\frac{1}{8}$ to $\frac{3}{8}$	0.87–0.79					
Heat-Absorbing Pattern	$\frac{1}{8}$	—					
Grey Sheet	$\frac{3}{16}, \frac{7}{32}$	0.74, 0.71					
Heat-Absorbing Plate/Float[d]	$\frac{3}{16}, \frac{1}{4}$	0.46	0.57	0.53	0.45	0.30	0.36
Heat-Absorbing Pattern	$\frac{1}{8}, \frac{1}{4}$	—					
Grey Sheet	$\frac{1}{8}, \frac{7}{32}$	0.59, 0.45					
Heat-Absorbing Plate/Float or Pattern	—	0.44–0.30	0.54	0.52	0.40	0.28	0.32
Heat-Absorbing Plate/Float[d]	$\frac{3}{8}$	0.34					
Heat-Absorbing Plate or Pattern	—	0.29–0.15	0.42	0.40	0.36	0.28	0.31
		0.24					
Reflective Coated Glass							
S.C.[c] = 0.30			0.25	0.23			
0.40			0.33	0.29			
0.50			0.42	0.38			
0.60			0.50	0.44			

[a] Refer to manufacturer's literature for values.
[b] For vertical blinds with opaque white and beige louvers in the tightly closed position, SC is 0.25 and 0.29 when used with glass of 0.71 to 0.80 transmittance.
[c] Shading Coefficient for glass with no shading device.
[d] Refers to grey, bronze, and green tinted heat-absorbing plate/float glass.

Reprinted by permission from "ASHRAE Handbook of Fundamentals" ASHRAE, New York, 1972.

Table 1-31. Shading Coefficients for Insulating Glass with Indoor Shading by Venetian Blinds and Roller Shades

Type of Glass	Nominal Thickness, each light	Solar Trans.[b] Outer Pane	Solar Trans.[b] Inner Pane	Venetian Blinds[c] Medium	Venetian Blinds[c] Light	Roller Shade Opaque Dark	Roller Shade Opaque White	Roller Shade Translucent Light
Regular Sheet Out	$\frac{3}{32}, \frac{1}{8}$	0.87	0.87					
Regular Sheet In				0.57	0.51	0.60	0.25	0.37
Regular Plate/Float Out	$\frac{1}{4}$	0.80	0.80					
Regular Plate/Float In								
Heat-Absorbing Plate/Float[d] Out	$\frac{1}{4}$	0.46	0.80	0.39	0.36	0.40	0.22	0.30
Regular Plate/Float In								
Reflective Coated Glass SC[e] = 0.20				0.19	0.18			
0.30				0.27	0.26			
0.40				0.34	0.33			

[a] Refers to factory-fabricated units with $\frac{3}{16}$, $\frac{1}{4}$, or $\frac{1}{2}$ in. air space, or to prime windows plus storm windows.
[b] Refer to manufacturer's literature for exact values.
[c] For vertical blinds with opaque white or beige louvers, tightly closed, SC is approximately the same as for opaque white roller shades.
[d] Refers to bronze or green tinted heat-absorbing plate/float glass.
[e] Shading Coefficient for glass with no shading device.

Reprinted by permission from "ASHRAE Handbook of Fundamentals," ASHRAE, New York, 1972.

Table 1-32. Shading Coefficients for Double Glazing with Between-glass Shading

Type of Glass	Nominal Thickness, each pane	Solar Trans.[a] Outer Pane	Solar Trans.[a] Inner Pane	Description of Air Space	Venetian Blinds Light	Venetian Blinds Medium	Louvered Sun Screen
Regular Sheet Out Regular Sheet In Regular Plate Out Regular Plate In	$\frac{3}{32}$, $\frac{1}{8}$ $\frac{1}{4}$	0.87 0.80	0.87 0.80	Shade in contact with glass or shade separated from glass by air space. Shade in contact with glass-voids filled with plastic.	0.33 —	0.36 —	0.43 0.49
Heat-Abs. Plate/Float[b] Out Regular Plate In	$\frac{1}{4}$	0.46	0.80	Shade in contact with glass or shade separated from glass by air space. Shade in contact with glass-voids filled with plastic.	0.28 —	0.30 —	0.37 0.41

[a] Refer to manufacturer's literature for exact values.
[b] Refers to grey, bronze, and green tinted heat-absorbing plate/float glass.
Reprinted by permission from "ASHRAE Handbook of Fundamentals," ASHRAE, New York 1972.

Table 1-33. Shading Coefficients for Hollow Glass Block Wall Panels[a]

Type of Glass Block[b]	Description of Glass Block	Shading Coefficient[c]	
		Panels[d] in the Sun	Panels[e] in the Shade (N, NW, W, SW)
Type I	Glass Colorless or Aqua Smooth Face A, D: Smooth B, C: Smooth or wide ribs, or flutes horizontal or vertical, or shallow configuration. E: None	0.65	0.40
Type IA	Same as Type I except A: Ceramic Enamel on exterior face.	0.27	0.20
Type II	Same as Type I except E: Glass fiber screen.	0.44	0.34
Type III	Glass Colorless or Aqua A, D: Narrow vertical ribs or flutes. B, C: Horizontal light-diffusing prisms, or horizontal light-directing prisms. E: Glass fiber screen.	0.33	0.27
Type IIIA	Same as Type III except, E: Glass fiber screen with green ceramic spray coating, or glass fiber screen and gray glass, or glass fiber screen with light-selecting prisms.	0.25	0.18

[a] For glass block used in horizontal skylights see Tables 28 and 29, Chapter 26 of the 1963 ASHRAE Guide And Data Book.

[b] All values are for 7¾ × 7¾ × 3⅞ in. block, set in light-colored mortar. For 11¾ × 11¾ × 3⅞ in. block increase coefficients by 15 percent, and for 5¾ × 5¾ × 3⅞ in. blocks reduce coefficients by 15 percent.

[c] Shading coefficients are to be applied to Heat Gain Factors for one hour earlier than the time for which the load calculation is made to allow for heat storage in the panel.

[d] Shading coefficients are for peak load condition, but provide a close approximation for other conditions. For more precise values for other conditions, see Reference 20.

[e] For NE, E, and SE panels in the shade add 50 percent to the values listed for panels in the shade.

Reprinted by permission from "ASHRAE Handbook of Fundamentals," ASHRAE, New York, 1972.

Table 1-34. Overall Coefficients of Heat Transmission
(U Values) for Fenestration under Summer Conditions
(7.5 mph Wind Outdoors, Still Air Indoors)

Type of Glass	U-Value No Shading	U-Value Internal Shading[a]
Any Uncoated Single Glass[c]	1.06	0.81
Insulating Glass,[c] $\frac{3}{16}$ in. Air Space	0.66	0.54
uncoated $\frac{1}{4}$ in. Air Space	0.65	0.52
$\frac{3}{8}$ in. Air Space	0.61	0.50
$\frac{1}{2}$ in. Air Space	0.59	0.48
Prime Window Plus Storm Window, Air Space 1 in. or more	0.54[b]	0.47[b]

	No Supplementary Shading
Double Glazing with Between-Glass Shading Louvered Sun Screen Separated by Air Space	0.63
Venetian Blinds, Closed, in Air Space	0.44
Glass Block Panels[d]	
Types I and II	0.56
Types II, III, and IIIA	0.48

[a] Values apply to tightly closed Venetian and vertical blinds, draperies, and roller shades.
[b] Values apply to storm sash with a tight air space. Air leakage present in virtually all storm windows will, in effect, increase this value.
[c] U-values can be substantially reduced by low-emittance coatings applied to the inner surface of single or double glazing and to an air-space surface of insulating glass. Consult manufacturers for applicable U-values.
[d] Values listed are for $7\frac{3}{4} \times 7\frac{3}{4} \times 3\frac{7}{8}$ in. block. For $11\frac{3}{4} \times 11\frac{3}{4} \times 3\frac{7}{8}$ in. block, reduce the listed value by 0.04, and for $5\frac{3}{4} \times 5\frac{3}{4} \times 3\frac{7}{8}$ in. block, increase the listed value by 0.04 See Table 1-33 for definition of types.

Reprinted by permission from "ASHRAE Handbook of Fundamentals," ASHRAE, New York, 1972.

Table 1-35. Rates of Heat Gain from Occupants of Conditioned Spaces[a]

Degree of Activity	Typical Application	Total Heat Adults, Male, Btu/Hr	Total Heat Adjusted,[b] Btu/Hr	Sensible Heat, Btu/Hr	Latent Heat, Btu/Hr
Seated at rest	Theater—Matinee	390	330	225	105
	Theater—Evening	390	350	245	105
Seated, very light work	Offices, hotels, apartments	450	400	245	155
Moderately active office work	Offices, hotels, apartments	475	450	250	200
Standing, light work; or walking slowly	Department store, retail store, dime store	550	450	250	200
Walking; seated Standing; walking slowly	Drug store, Bank	550	500	250	250
Sedentary work	Restaurant[c]	490	550	275	275
Light bench work	Factory	800	750	275	475
Moderate dancing	Dance hall	900	850	305	545
Walking 3 mph; moderately heavy work	Factory	1000	1000	375	625
Bowling[d] Heavy work	Bowling alley Factory	1500	1450	580	870

[a] *Note:* Tabulated values are based on 75 F room dry-bulb temperature. For 80 F room dry-bulb, the total heat remains the same, but the sensible heat values should be decreased by approximately 20 percent, and the latent heat values increased accordingly.

[b] *Adjusted total heat gain* is based on normal percentage of men, women, and children for the application listed, with the postulate that the gain from an adult female is 85 percent of that for an adult male, and that the gain from a child is 75 percent of that for an adult male.

[c] Adjusted total heat value for *sedentary work, restaurant,* includes 60 Btu per hour for food per individual (30 Btu sensible and 30 Btu latent).

[d] For *bowling* figure one person per alley actually bowling, and all others as sitting (400 Btu per hour) or standing (550 Btu per hour).

Reprinted by permission from ASHRAE "Handbook of Fundamentals," ASHRAE, New York, 1972.

Table 1-36. Internal Heat Gain from Miscellaneous Appliances

Appliance	Manufacturer's rating		Recommended rate of heat gain, Btu/hr		
	Watts	Btu/hr	Sensible	Latent	Total
Electrical Appliances					
Hair dryer:					
Blower type............................	1,580	5,400	2,300	400	2,700
Helmet type............................	705	2,400	1,870	330	2,200
Permanent-wave machine, 60 heaters at 25 w,					
36 in normal use.........................	1,500	5,000	850	150	1,000
Neon sign, per linear foot of tube:					
½″ diam................................	.....	.....	30	.....	30
⅜″ diam................................	.....	.....	60	.....	60
Sterilizer, instrument........................	1,100	3,750	650	1,200	1,850
Gas-burning Appliances					
Lab burners:					
Bunsen, ⁷⁄₁₆″ barrel......................	.....	3,000	1,680	420	2,100
Fishtail, 1½″ wide.......................	.....	5,000	2,800	700	3,500
Meeker, 1″ diam.........................	.....	6,000	3,360	840	4,200
Gaslight, per burner, mantle type...........	.....	2,000	1,800	200	2,000
Cigar lighter, continuous flame...............	.....	2,500	900	100	1,000

Reprinted by permission from "ASHRAE Handbook of Fundamentals," ASHRAE, New York, 1972.

Table 1-37. Heat Gain from Electric Motors

Nameplate rating of motor, hp	Average motor efficiency in continuous operation	Btu/hr to room air per rated hp of motor		
		Motor outside of room, driven device inside room	Motor in room, driven device outside of room	Motor and driven device both inside room
⅙–½	0.60	2546	1700	4246
½–3	.69	2546	1100	3646
3–20	.85	2546	400	2946

General rule for motors: if H_m = Btu/hr of motor input,

$$H_m = \frac{2{,}546 \times hp \ (\text{connected load})}{\text{motor efficiency}}$$

NOTE: Where possible obtain actual value of motor efficiency. Where not possible: for motors use average efficiencies as listed above; for motor generators use average efficiency sets up to 3 hp. as 0.55; for larger sets use average efficiency as 0.80.

SOURCE: Strock and Koral, "Handbook of Air Conditioning, Heating, and Ventilating," The Industrial Press, New York, 1965.

Table 1-38. Electric Motor-driven Appliances

(Motor and Driven Appliance Both in Same Room)

Fans (blade diameters, in.)	Btu/hr	Appliances	Btu/hr
Ceiling 32	340	Clock..........................	7
52	410	Hair dryer.....................	1900
56	600	Drink mixer....................	240
Desk or wall 8	120	Sewing machine (domestic).......	220
10	140	Vacuum cleaner (domestic).......	250
12	200	Hair clipper...................	78
16	300	Vibrator (beauty).............	11

NOTE: Figures are thermal equivalents of nameplate rating, corrected for motor efficiency.	NOTE: Figures are thermal equivalents of nameplate ratings.

SOURCE: Strock and Koral, "Handbook of Air Conditioning, Heating, and Ventilating," The Industrial Press, New York, 1965.

Table 1-39. Electric Refrigerators

With well insulated cabinet in 80F air and 40F inside cabinet

Electric Refrigerators

Electric motor and air-cooled condenser in cabinet in room air		Cold Cabinet in Room (compressor and condenser remote) With well-insulated cabinet in 80°F air and 40°F inside cabinet allow a cooling effect as follows:	
Cabinet volume, ft³	Btu/hr (thermal equivalent of motor input)	Cabinet volume, ft³	Btu/(hr)(ft³) of cabinet volume
2–4	530	2–4	100–75
5	710	5–6	70–65
6–10	850	7–10	60–55
12–18	1060	12, 14, 16	55–50
		20, 25, 30	50, 45, 40

SOURCE: Strock and Koral, "Handbook of Air Conditioning, Heating, and Ventilating," The Industrial Press, New York, 1965.

Table 1-40. Recommended Inside Design Conditions—Summer

Type of application	Summer				
	Deluxe		Commercial practice		
	Dry-bulb, °F	Rel. hum., %	Dry-bulb, °F	Rel. hum., %	Temp. swing,* °F
General Comfort Apartment, house, hotel, office, hospital, school, etc.	74–76	50–45	77–79	50–45	2–4
Retail Shops (Short-term occupancy) Bank, barber, or beauty shop, department store, supermarket, etc............................	76–78	50–45	78–80	50–45	2–4
Low Sensible Heat Factor Applications (High latent load) Auditorium, church, bar, restaurant, kitchen, etc...	76–78	55–50	78–80	60–50	1–2
Factory Comfort Assembly areas, machining rooms, etc.............	77–80	55–45	80–85	60–50	3–6

* Temperature swing is above the thermostat setting at peak summer load conditions.
SOURCE: "Carrier Corporation System Design Manual," Part I, Load Estimating, 1970.

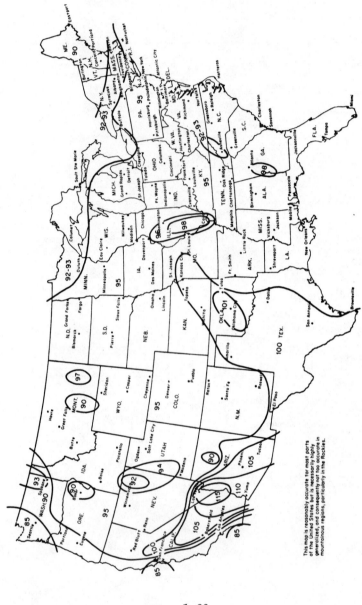

This map is reasonably accurate for most parts of the United States but is necessarily highly generalized, and consequently not too accurate in mountainous regions, particularly in the Rockies.

FIG. 1-12. Summer outside dry-bulb design temperature, °F. (Strock and Koral, "Handbook of Air Conditioning, Heating, and Ventilating," The Industrial Press, New York, 1965.)

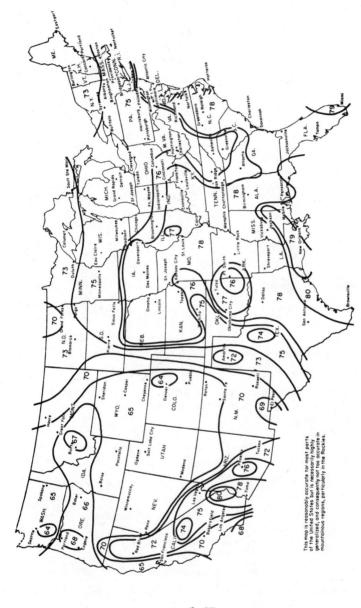

This map is reasonably accurate for most parts of the United States but is necessarily highly generalized, and consequently not too accurate in mountainous regions, particularly in the Rockies.

FIG. 1-13. Summer outside wet-bulb design temperature, °F. (Strock and Koral, "Handbook of Air Conditioning, Heating, and Ventilating," 2d ed., 1965.)

Supply Air Temperature. The room cooling load is the sum of external and internal sensible and latent heat gains plus the difference in enthalpy between outside and room air for that portion of outside air that does not contact the cooling-coil surfaces. The percentage of air that passes through a cooling coil untreated is the numerical value of the *coil bypass factor;* e.g., a bypass factor of 20 per cent represents a cooling-coil saturation efficiency of 80 per cent.

The ratio of room sensible heat gains to total room sensible and latent heat gains is the room *sensible heat ratio* (RSHR).

$$\text{RSHR} = Q_{rs}/(Q_{rs} + Q_{ri}) \qquad (1\text{-}27)$$

It represents the ratio of sensible cooling capacity to the total cooling capacity required of the supply air to satisfy room conditions. It is used to plot the slope of the *room-condition line* on a psychrometric chart (Fig. 1-14) for the determination of the *apparatus dew point* (ADP).

The actual supply air temperature and off coil wet-bulb temperature will depend on the bypass characteristic of the selected cooling coil (Fig. 1-15).

Supply Air Rate. The rate of supply air required is expressed by

$$Q_{sa} = Q_{rs}/1.08(t_r - t_s) \qquad (1\text{-}28)$$

where Q_{sa} = supply air, cfm

   t_r = room design temperature, °F

   t_s = supply air temperature, °F

   1.08 = (60 min)[0.244 Btu/(lb)(°F)](0.075 lb/ft³) $\qquad (1\text{-}29)$

4-6. Air Distribution

Outlets. *Purpose.* Outlets are designed:

1. To control air motion, noise level, and temperature gradients caused by the introduction of air to and the removal of air from a space

2. To counteract the natural convection and radiation effects within the room

Supply Outlets. Supply outlets should be selected on the basis of manufacturers' data. Factors which usually affect the selection of supply outlets are (1) noise, (2) location of outlet, (3) temperature of supply air, and (4) area of diffusion.

Return Outlets. Selection of return registers or grilles is usually governed by face velocity.

Table 1-41. Recommended Return Intake Face Velocities

Intake Location	Velocity over Gross Area, FPM
Above occupied zone	800 up
Within occupied zone, not near seats	600–800
Within occupied zone, near seats	400–600
Door or wall louvers	200–300
Undercutting of doors (through undercut area)	200–300

Reprinted by permission from "ASHRAE Handbook of Fundamentals," ASHRAE, New York, 1972.

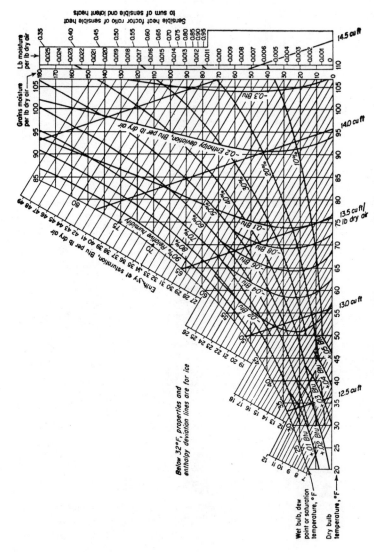

FIG. 1-14. Psychrometric chart—normal temperatures.

1–69

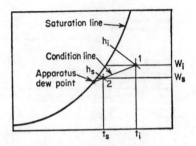

FIG. 1-15. Apparatus dew-point and condition line.

Ductwork. *Air Velocity.* Supply and return air ducts and apparatus are sized on the basis of air quantity, within the limitations of allowable friction losses, velocity, and noise (Table 1-42).

Table 1-42. Recommended and Maximum Duct Velocities for Conventional Systems

Designation	Residences	Schools, theaters, public buildings	Industrial buildings
Recommended velocities, fpm			
Outdoor air intakes*	500	500	500
Filters*	250	300	350
Heating coils*	450	500	600
Air washers	500	500	500
Fan outlets	1,000–1,600	1,300–2,000	1,600–2,400
Main ducts	700–900	1,000–1,300	1,200–1,800
Branch ducts	600	600–900	800–1,000
Branch risers	500	600–700	800
Maximum velocities, fpm			
Outdoor air intakes*	800	900	1,200
Filters*	300	350	350
Heating coils*	500	600	700
Air washers	500	500	500
Fan outlets	1,700	1,500–2,200	1,700–2,800
Main ducts	800–1,200	1,100–1,600	1,300–2,200
Branch ducts	700–1,000	800–1,300	1,000–1,800
Branch risers	650–800	800–1,200	1,000–1,600

* These velocities are for total face area, not the net free area; other velocities in the table are for net free area.
SOURCE: "ASHRAE Guide and Data Book," chap. 12, table 6, ASHRAE, New York, 1963.

High-velocity air distribution (2,000 to 6,000 fpm) using much smaller ducts and operating at greater pressures is used when space is critical.

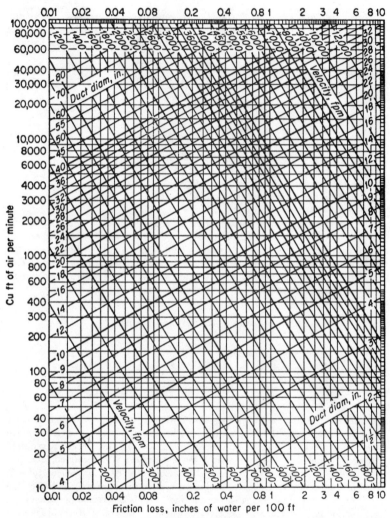

Fig. 1-16. Friction loss for usual air conditions. This chart applies to smooth round galvanized-iron ducts, and is based on air at 70°F and 29.96 in. Hg abs pressure. For air of different density the friction may be assumed to vary directly with the density.

Pressure Losses in Duct Systems.   Pressure losses in duct systems are due to friction of the air in contact with the sides of the duct and dynamic losses caused by changes of duct shape or direction and by obstructions to flow.

Friction:
$$H_f = f \frac{L}{D} \left(\frac{V}{4005} \right)^2$$
(1-30)

where H_f = head loss due to friction, in. H_2O

L = length of duct, ft

D = diameter of duct, ft

V = velocity of air, fpm

f = nondimensional friction coefficient

Dynamic losses: $H_v = CV^2/4,005$

where H_v = velocity-head loss, in. H_2O

C = experimentally determined constant

V = air velocity, fpm

Design Methods.   The *equal-friction method* is applicable primarily to systems using low or moderate velocities where the velocity head is not an important factor.   A friction drop per 100 ft of length is chosen, and the duct mains and branches are all sized on the basis of this friction drop.   This will invariably result in higher velocities in the mains, where they can be tolerated, and low velocities in the branches, where they are desirable.

Table 1-43. Friction Drops

Application	Friction Drop, In. H_2O/100 Ft
Noise critical, low velocity	0.05–0.07
Average application	0.08–0.1
Equipment rooms, industrial applications	0.11–0.13

The *static-regain method* is used for both conventional and high-velocity systems.   It is especially applicable in the latter, where the velocity head may be appreciable.   In the static-regain method, the static pressure required to give proper air flow through the system outlets is determined, and this pressure is maintained by reducing the velocity at each branch or takeoff, so that the recovery in pressure due to reduction of velocity balances the friction loss in the preceding section of duct.   This is possible because of the convertibility of static and velocity pressures.   For practical applications it is usually assumed that 50 per cent of the velocity pressure available will be converted to static pressure.

$$H_R = 0.5 \left(\frac{V_1}{4005} \right)^2 - \left(\frac{V_2}{4005} \right)^2$$
(1-31)

where H_R = head recovered, in. H_2O

V_1 = system inlet velocity, fpm

V_2 = system outlet velocity, fpm

Fans. *Fan Laws*

Quantity required	Cfm	Total head delivered by wheel	Rpm	Hp	Wheel diam*
Cfm............................	....	$H_t^{1/2}$	rpm†	hp$^{1/3}$	D
Total head delivered by wheel.....	cfm^2	.....	rpm^2	hp$^{2/3}$	D^2
Rpm............................	cfm	$H_t^{1/2}$	.....	hp$^{1/3}$	D^2
Hp............................	cfm^3	$H_t^{3/2}$	rpm^3	....	D^3

* Constant speed.
† Constant head.

Equations

Mechanical efficiency $= \dfrac{0.0001575 \times \text{cfm} \times \text{total pressure, in. H}_2\text{O}}{\text{horsepower input}}$ (1-32)

Equation (1-32) is applicable to fans operating with high outlet velocity pressure relative to static pressure.

Static efficiency $= \dfrac{0.0001573 \times \text{cfm} \times \text{static pressure, in. H}_2\text{O}}{\text{horsepower input}}$ (1-33)

Equation (1-33) is more applicable to fans with high static pressure relative to velocity pressure.

Characteristics

Table 1-44. Relative Characteristics of Centrifugal Fans

Characteristic	Backward	Radial	Forward
First cost................	High	Medium	Low
Efficiency...............	High	Medium	Poor
Stability of operation......	Good	Good	Poor
Space required............	Medium	Medium	Small
Tip speed...............	High	Medium	Low
Resistance to abrasion.....	Medium	Good	Poor

Table 1-45. Outlet Velocities for Optimum Performance
of Typical Ventilating Fans

Static pressure, in. water	Centrifugal fans— outlet velocity, fpm	Tube-axial and vane-axial fans— outlet velocity at wheel diam., fpm
¼	400–1,100	950–1,500
½	550–1,300	1,350–1,900
¾	700–1,500	1,650–2,350
1	800–1,750	1,900–2,700
1½	1,000–2,450	2,350–3,300
2	1,150–2,800	2,700–3,800
2½	1,250–3,200	3,000–4,300
3	1,400–3,500	3,300–4,700
4	1,600–4,000	
6	2,000–4,900	
8	2,300–5,650	
10	2,500–6,300	

SOURCE: "ASHRAE Guide and Data Book," chap. 40, fig. 1, ASHRAE, New York, 1963.

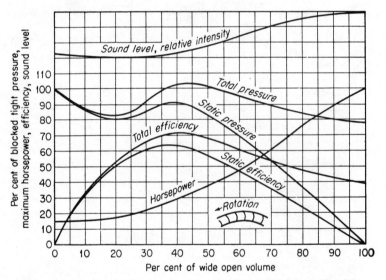

FIG. 1-17. Percentage performance curves of a forward-blade centrifugal fan.

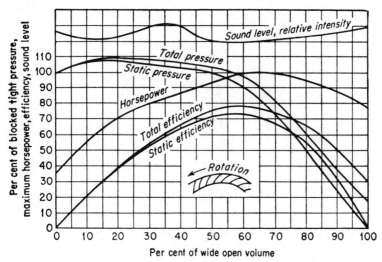

Fig. 1-18. Percentage performance curves of a backward-curved-blade centrifugal fan.

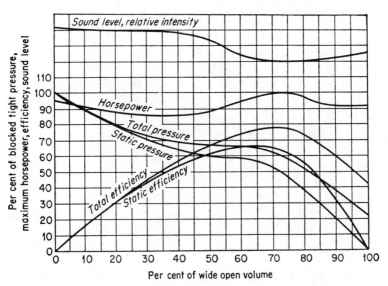

Fig. 1-19. Percentage performance curves of an axial-flow fan.

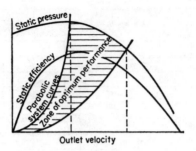

FIG. 1-20. Zone of optimum performance for fans.

Correction Factors for Temperature and Altitude

Table 1-46. Correction Factor for Altitude and Temperature to Air Volume

Altitude, ft above sea level....	0	1,000	2,000	3,000	4,000	5,000	6,000	7,000	8,000
Barometric pressure, in. Hg...	29.92	28.86	27.82	26.81	25.84	24.89	23.98	23.09	22.22
Air temp, °F	Correction factors								
70	1.040	1.003	0.967	0.932	0.898	0.865	0.833	0.803	0.772
100	0.984	0.948	.915	.882	.850	.818	.788	.759	.731
150	.904	.872	.840	.801	.781	.752	.724	.698	.672
200	.835	.805	.777	.749	.722	.694	.668	.645	.620
250	.777	.749	.722	.696	.671	.647	.622	.599	.577
300	.725	.699	.674	.649	.628	.603	.580	.560	.538
350	.680	.656	.632	.609	.588	.566	.545	.525	.505
400	.641	.618	.596	.574	.553	.533	.512	.495	.476
450	.605	.583	.564	.543	.523	.503	.485	.467	.450
500	.574	.553	.534	.515	.496	.477	.460	.443	.426
550	.546	.526	.508	.490	.472	.454	.438	.421	.406
600	.520	.501	.484	.466	.449	.433	.416	.401	.387
650	.496	.478	.462	.444	.428	.413	.397	.383	.368
700	.475	.458	.442	.426	.411	.395	.381	.367	.354

NOTE: Equivalent cfm = $\dfrac{\text{cfm at actual conditions}}{\text{correction factor}}$

SOURCE: "Bulletin 3576-B, Correction Factors for Temperature and Altitude," Buffalo Forge Co., Buffalo, N.Y.

1-7. Refrigeration

Refrigeration cycles are discussed in Sec. 8. Thermodynamic data for typical refrigerants are given in Sec. 3.

1-8. Water Distribution

Chilled-water Systems

Temperature differential

Application	Temperature rise, °F
Close-coupled system on one floor	5–8
Two- or three-story building	8–11
Multistory building	12–20

$$Gpm = \frac{\text{total load Btu/hr} + \text{piping heat gains} + \text{pump heat}}{500 \times \text{temperature differential}} \quad (1\text{-}34)$$

Condenser Water Systems. For electrically driven refrigeration compressors a temperature differential of 10°F may be assumed, and for steam-driven equipment a temperature differential of 20°F is usual. In the latter case the refrigeration and steam condensers are piped in series with a temperature rise of approximately 10°F each.

Table 1-47. Heat Rejection of Typical Processes

Equipment	Btu/min/ton	Btu/kwhr	Btu/bhp-hr
Refrigeration compressors, open drive	250		
Refrigeration compressors, hermetic	300		
Refrigeration absorption system	550		
Steam jet refrigerating system	550		
Steam electric power plant, kw:			
500	...	11,210	
1,000	...	10,750	
5,000	...	8,150	
7,500	...	7,700	
10,000	...	7,020	
Diesel engine jacket and lube oil:			
Four-cycle, supercharged	...	.....	2,600
Four-cycle, nonsupercharged	...	.....	3,000
Two-cycle, crank-case compressor	...	.....	2,000
Two-cycle, pump-scavenging (large unit)	...	.....	2,500
Two-cycle, pump-scavenging (high-speed)	...	.....	2,200
Natural-gas engine:			
Four-cycle	...	.....	4,500
Two-cycle	...	.....	4,000

SOURCE: "ASHRAE Guide and Data Book," chap. 37, table 4, ASHRAE, New York, 1961.

Atmospheric Cooling Equipment. The lowest temperature to which water can be cooled in atmospheric cooling equipment is the wet-bulb temperature of the ambient air.

Water Cooling Effectiveness in Per Cent

$$E = \frac{(\text{hot-water temperature} - \text{cold-water temperature}) \times 100}{\text{hot-water temperature} - \text{wet-bulb temperature of entering air}} \quad (1\text{-}35)$$

The cold-water temperature must be chosen to place the requirement within the effectiveness range of the equipment used.

Table 1-48. **Effectiveness of Water Cooling Equipment**

Cooling equipment	Water cooling effectiveness, %		
	Minimum	Typical	Maximum
Spray ponds......................	30	40–50	68
Spray-filled atmospheric towers......	40	45–55	60
Atmospheric deck towers............	50	50–60	90
Mechanical draft towers............	50	55–75	93

source: "Heating, Ventilating, and Air Conditioning Guide," chap. 34, table 3, ASHRAE, New York, 1958

Makeup Water. Makeup water is introduced to replace losses due to evaporation, drift, and blowdown.

If all water were cooled by evaporation, the loss by evaporation for the usual 10°F cooling range would be

$$\text{Evaporation \%} = \frac{Q \times 100}{8.3 \times \text{gpm} \times h_{fg}} \qquad (1\text{-}36)$$

where Q = total heat rejected, Btu/hr

gpm = total condenser water circulated, gpm

h_{fg} = evaporation heat of water, Btu/lb, at ambient design temperature

In practice, the loss of circulating water by evaporation due to additional cooling by sensible heat transfer will vary from about 0.64 per cent in winter to 0.88 per cent in the summer for a water-cooling range of 10°F.

Drift losses depend on the tower design, but generally, from the cooling tower, they are limited to 0.2 per cent of the circulated rate.

The makeup water replacing losses due to evaporation, drift, and blowdown introduces dissolved solids into the system.

To prevent excessive concentration, a portion of the circulating water is wasted. The quantity of blowdown depends on the original quantity of dissolved solids in the makeup water and the permissible concentration.

For larger installation, chemical water-treatment processes are used, which also require a controlled blowdown rate.

1-9. Pumps

The performance of pumps is discussed in Sec. 8. Methods for calculating pressure drop in pipe and fittings are presented in Sec. 2.

1-10. Drainage

Sanitary Load

Table 1-49. Drainage Fixture Unit Values for
Various Plumbing Fixtures

Type of Fixture or Group of Fixtures	Drainage Fixture Unit Value (d.f.u.)
Sinks:	
Surgeon's	3
Flushing rim (with valve)	6
Service (trap standard)	3
Service (P trap)	2
Pot, scullery, etc.*	4
Urinal, pedestal, syphon jet blowout	6
wall lip	4
stall, washout	4
Urinal trough (each 6-ft section)	2
Wash sink (circular or multiple) each set of faucets	2
Water closet, tank-operated	4
valve-operated	6
Fixtures not listed above:	
Trap Size 1¼ in. or less	1
1½ in.	2
2 in.	3
2½ in.	4
3 in.	5
4 in.	6

* See Sec. 11.4.2 for method of computing equivalent fixture unit values for devices or equipment which discharge continuous or semicontinuous flows into sanitary drainage systems.
SOURCE: "National Standard Plumbing Code," Table 11.4.1, 1973.

Table 1-50. Size of Nonintegral Traps for
Different Plumbing Fixtures

Plumbing fixture	Trap size, in.
Dental lavatory	1¼
Drinking fountain	1¼
Dishwasher, commercial	2
domestic (nonintegral trap)	1½
Floor drain	2
Food waste grinder, commercial use	2
domestic use	1½
Kitchen sink, domestic, with food waste grinder unit	1½
domestic	1½
domestic, with dishwasher	1½
Lavatory, common	1¼
barber shop, beauty parlor, or surgeon's	1½
multiple type (wash fountain or wash sink)	1½
Laundry tray (1 or 2 compartments)	1½
Shower stall or drain	2
Sinks:	
Surgeon's	1½
Flushing rim type, flush valve supplied	3
Service type with floor outlet trap standard	3
Service trap with P trap	2
Commercial (pot, scullery, or similar type)	2
Commercial (with food grinder unit)	2

* Separate trap required for wash tray and separate trap required for sink compartment with food waste grinder unit.
SOURCE: "National Standard Plumbing Code," Table 5.2, 1973.

Table 1-51. Size and Length of Vents

Size of soil or waste stack, in.	Fixture units connected	Diameter of vent required, in.								
		1¼	1½	2	2½	3	4	5	6	8
		Maximum length of ven, ft								
1½	8	50	150							
1½	10	30	100							
2	12	30	75	200						
2	20	26	50	150						
2½	42	...	30	100	300					
3	10	...	30	100	100	600				
3	30	...	...	60	200	500				
3	60	...	...	50	80	400				
4	100	...	...	35	100	260	1,000			
4	200	...	...	30	90	250	900			
4	500	...	...	20	70	180	700			
5	200	...	...	...	35	80	350	1,000		
5	500	...	...	...	30	70	300	900		
5	1,100	...	...	...	20	50	200	700		
6	350	...	...	...	25	50	200	400	1,300	
6	620	...	...	...	15	30	125	300	1,100	
6	960	...	...	...	...	24	100	250	1,000	
6	1,900	...	...	...	...	20	70	200	700	
8	600	...	...	...	...	...	50	150	500	1,300
8	1,400	...	...	...	...	...	40	100	400	1,200
8	2,200	...	...	...	...	...	30	80	350	1,100
8	3,600	...	...	...	...	...	25	60	250	800
10	1,000	...	...	...	...	...	.....	75	125	1,000
10	2,500	...	...	...	...	...	.....	50	100	500
10	3,800	...	...	...	...	...	.....	30	80	350
10	5,600	...	...	...	...	...	.....	25	60	250

SOURCE: "National Standard Plumbing Code," Table 12.16.6, 1973.

Table 1-52. Maximum Length of Trap Arm

Size of fixture drain, in.	Distance—trap to vent
1¼	2 ft 6 in
1½	3 ft 6 in
2	5 ft
3	6 ft
4	10 ft

SOURCE: 'National Standard Plumbing Code," Table 12.8.1, 1973.

Storm-water Load

$$S = ARC/96 \qquad (1\text{-}37)$$

where S = storm-water quantity, gpm
A = area being drained, ft^2
R = design rate of rainfall, in./hr
C = ratio of runoff to rainfall

Design rate of rainfall varies with locality but is usually between 3 and 6 in./hr.

Table 1-53. Runoff Coefficients for Rational Formula

Type of area	Flat: slope <2%	Rolling: slope 2–10%	Hilly: slope >10%
Pavements, roofs, etc.	0.90	0.90	0.90
City business areas	.80	.85	.85
Suburban residential areas	.45	.50	.55
Dense residential areas	.60	.65	.70
Grassed areas	.25	.30	.30
Earth areas	.60	.65	.70
Cultivated land:			
Impermeable (clay, loam)	.50	.55	.60
Permeable (sand)	.25	.30	.35
Meadows and pasture lands	.25	.30	.35
Forests and wooded areas	.10	.15	.20

Pipe Sizing

Table 1-54. Building Drains and Sewers

Pipe diameter, in.	Maximum number of fixture units that may be connected to any portion of the building drain or the building sewer including branches of the building drain			
	Fall per foot			
	1/16 in.	1/8 in.	1/4 in.	1/2 in.
2	.....	.....	21	26
2½	.....	.....	24	31
3	.....	36*	42*	50*
4	.....	180	216	250
5	.....	390	480	575
6	.....	700	840	1,000
8	1,400	1,600	1,920	2,300
10	2,500	2,900	3,500	4,200
12	3,900	4,600	5,600	6,700
15	7,000	8,300	10,000	12,000

* Not over two water closets or two bathroom groups.
NOTE: On-site sewers that serve more than one building may be sized according to the current standards and specifications of the Administrative Authority for public sewers.
SOURCE: "National Standard Plumbing Code," Table 11.5.1A, 1973.

Size of Combined Drains and Sewers. For combined storm and sanitary systems, drain sizing is based on fixture units and the storm drainage area is converted to equivalent fixture units.

Where the total fixture-unit load on the combined drain is less than 256 fixture units, the equivalent drainage area in horizontal projection is taken as 1,000 ft².

When the total fixture-unit load exceeds 256 fixture units, each fixture unit is considered the equivalent of 3.9 ft² of drainage area.

If the rainfall to be provided for is more or less than 4 in./hr, the 1,000-ft² equivalent and the 3.9 ft² are adjusted by multiplying by 4 and dividing by the rainfall in inches per hour to be provided for.

Table 1-55. Size of Roof Gutters*

Diameter of gutter, in.†	Maximum projected roof area for gutters, $\frac{1}{16}$-in. slope‡	
	ft²	gpm
3	170	7
4	360	15
5	625	26
6	960	40
7	1,380	57
8	1,990	83
10	3,600	150

* Table 1-55 is based on a maximum rate of rainfall of 4 in./hr for a 5-min duration and 10-year return period. Where maximum rates are more or less than 4 in./hr, the figures for drainage area shall be adjusted by multiplying by 4 and dividing by the local rate in inches per hour.
† Gutters other than semicircular may be used provided they have an equivalent cross-sectional area.
‡ Capacities given for slope of $\frac{1}{16}$ in./ft shall be used when designing for greater slopes.
SOURCE: "National Standard Plumbing Code." Table 13.6.3, 1973.

Table 1-56. Size of Vertical Conductors and Leaders*

Size of leader or conductor, in.†	Maximum projected roof area	
	ft²	gpm
2	544	23
2½	987	41
3	1,610	67
4	3,460	144
5	6,280	261
6	10,200	424
8	22,000	913

* Table 1-56 is based on a maximum rate of rainfall of 4 in./hr and on the hydraulic capacities of vertical circular pipes flowing between one-third and one-half full at terminal velocity, computed by the method of NBS Mono. 31. Where maximum rates are more or less than 4 in./hr, the figures for drainage area shall be adjusted by multiplying by 4 and dividing by the local rate in inches per hour.
† The area of rectangular leaders shall be equivalent to that of the circular leader or conductor required. The ratio of width to depth of rectangular leaders shall not exceed 3 to 1.
SOURCE: "National Standard Plumbing Code," Table 13.6.1, 1973.

Table 1-57. Horizontal Fixture Branches and Stacks

Pipe diameter, in.	Maximum number of fixture units that may be connected to			
	Any horizontal fixture branch*	Stack sizing for 3 stories or 3 intervals	Stack sizing for more than 3 stories	
			Total for stack	Total at story or branch interval
1½	3	4	8	2
2	6	10	24	6
2½	12	20	42	9
3	20†	48†	72†	20†
4	160	240	500	90
5	360	540	1,100	200
6	620	960	1,900	350
8	1,400	2,200	3,600	600
10	2,500	3,800	5,600	1,000
12	3,900	6,000	8,400	1,500
15	7,000			

* Does not include branches of the building drain.
† Not more than two water closets or bathroom groups within each branch interval nor more than six water closets or bathroom groups on the stack.
‡ Stacks shall be sized according to the total accumulated connected load at each story or branch interval and may be reduced in size as this load decreases to a minimum diameter of half of the largest size required.
SOURCE: "National Standard Plumbing Code," Table 11.5.1B, 1973.

Table 1-58. Size of Horizontal Storm Drains*

Drain diameter, in.	Maximum projected area for drains of various slopes					
	⅛-in. slope		¼-in. slope		½-in. slope	
	ft²	gpm	ft²	gpm	ft²	gpm
3	822	34	1,160	48	1,644	68
4	1,880	78	2,650	110	3,760	156
5	3,340	139	4,720	196	6,680	278
6	5,350	222	7,550	314	10,700	445
8	11,500	478	16,300	677	23,000	956
10	20,700	860	29,200	1,214	41,400	1,721
12	33,300	1,384	47,000	1,953	66,600	2,768
15	59,500	2,473	84,000	3,491	119,000	4,946

* Table 1-58 is based on a maximum rate of rainfall of 4 in./hr. Where maximum rates are more or less than 4 in./hr, the figures for drainage area shall be adjusted by multiplying by 4 and dividing by the local rate in inches per hour.
SOURCE: "National Standard Plumbing Code," Table 13.6.2, 1973.

1-11. Cold Water

Table 1-59. Estimating Demand

Supply systems predominantly for flush tanks		Supply systems predominantly for flush valves	
Load, water supply fixture units	Demand gpm	Load, water supply fixture units	Demand gpm
6	5		
8	6.5		
10	8	10	27
12	9.2	12	28.6
14	10.4	14	30.2
16	11.6	16	31.8
18	12.8	18	33.4
20	14	20	35
25	17	25	38
30	20	30	41
35	22.5	35	43.8
40	24.8	40	46.5
45	27	45	49
50	29	50	51.5
60	32	60	55
70	35	70	58.5
80	38	80	62
90	41	90	64.8
100	43.5	100	67.5
120	48	120	72.5
140	52.5	140	77.5
160	57	160	82.5
180	61	180	87
200	65	200	91.5
225	70	225	97
250	75	250	101
275	80	275	105.5
300	85	300	110
400	105	400	126
500	125	500	142
750	170	750	178
1,000	208	1,000	208
1,250	240	1,250	240
1,500	267	1,500	267
1,750	294	1,750	294
2,000	321	2,000	321
2,250	348	2,250	348
2,500	375	2,500	375
2,750	402	2,750	402
3,000	432	3,000	432
4,000	525	4,000	525
5,000	593	5,000	593
6,000	643	6,000	643
7,000	685	7,000	685
9,000	718	8,000	718
8,000	745	9,000	745
10,000	769	10,000	769

SOURCE: "National Standard Plumbing Code," Table 10.13.2B, 1973.

Table 1-60. Water Consumption per Capita

Occupancy	Gal as stated or gpcpd	Occupancy	Gal as stated or gpcpd
Office buildings.............	27–45	Laundries, per pound.......	3–5.7
Grade schools..............	5–10	Hotels, per room..........	300–525
High schools...............	15–20	Hospitals, per bed.........	125–350
Restaurants, per meal.......	0.5–4		

Table 1-61. Sizing the Water Supply System*

Fixture	Occupancy	Type of supply control	Load in fixture units†
Bathroom group‡........	Private	Flush valve for closet	8
Bathroom group‡........	Private	Flush tank for closet	6
Bathtub...............	Private	Faucet	2
Bathtub...............	Public	Faucet	4
Clothes washer.........	Private	Faucet	2
Clothes washer.........	Public	Faucet	4
Combination fixture.....	Private	Faucet	3
Kitchen sink............	Private	Faucet	2
Kitchen sink............	Hotel, restaurant	Faucet	4
Laundry trays (1 to 3)...	Private	Faucet	3
Lavatory...............	Private	Faucet	1
Lavatory...............	Public	Faucet	2
Separate shower........	Private	Mixing valve	2
Service sink............	Office, etc.	Faucet	3
Shower head...........	Private	Mixing valve	2
Shower head...........	Public	Mixing valve	4
Urinal, pedestal	Public	Flush valve	10
stall or wall......	Public	Flush valve	5
stall or wall......	Public	Flush tank	3
Water closet...........	Private	Flush valve	6
Water closet...........	Private	Flush tank	3
Water closet...........	Public	Flush valve	10
Water closet...........	Public	Flush tank	5

Water supply outlets for items not listed above shall be computed at their maximum demand, but in no case less than:

Fixture, in.	Number of fixture units	
	Private use	Public use
3/8	1	2
1/2	2	4
3/4	3	6
1	6	10

* For supply outlets likely to impose continuous demands, estimate continuous supply separately and add to total demand for fixtures.

† The given weights are for total demand. For fixtures with both hot and cold water supplies, the weights for maximum separate demands may be taken as 3/4 the listed demand for the supply.

‡ A bathroom group for the purposes of this table consists of not more than one water closet, one lavatory, one bathtub, one shower stall or not more than one water closet, two lavatories, one bathtub or one separate shower stall.

SOURCE: "National Standard Plumbing Code," Table 10.13.2.A, 1973.

Table 1-62. Proper Flow and Pressure Required During Flow for Different Fixtures

Fixture	Flow pressure*	Flow, gpm
Ordinary basin faucet..................	8	3.0
Self-closing basin faucet..............	12	2.5
Sink faucet, ⅜ in......................	10	3.5
Sink faucet, ½ in......................	5	4.5
Dishwasher..........................	15–25	†
Bathtub faucet.......................	5	6.0
Laundry tub cock, ¼ in...............	5	5.0
Shower..............................	12	3–10
Ball-cock for closet....................	15	3.0
Flush valve for closet.................	10–20	15–40‡
Flush valve for urinal.................	15	15.0
Garden hose, 50 ft, and sill cock........	30	5.0

* Flow pressure is the pressure psig in the pipe at the entrance to the particular fixture considered.
† Varies, see manufacturers' data.
‡ Wide range due to variation in design and type of flush-valve closets.
Reprinted by permission from "ASHRAE Handbook of Fundamentals," ASHRAE, New York, 1972.

1-12. Hot Water

Table 1-63. Maximum Daily (24-hr) Requirements for Hot Water in Gallons

Apartments and private homes with no. of rooms	Number of bathrooms				
	1	2	3	4	5
1	60				
2	70				
3	80				
4	90	120			
5	100	140			
6	120	160	200		
7	140	180	220		
8	160	200	240	250	
9	180	220	260	275	
10	200	240	280	300	
11	...	260	300	340	
12	...	280	325	380	450
13	...	300	350	420	500
14	...	...	375	460	550
15	...	...	400	500	600
16	...	...	...	540	650
17	...	...	...	580	700
18	...	...	...	620	750
19	...	...	...	...	800
20	...	...	...	...	850

Hotels:
Room with basin.............................. 10
Room with bath—transient..................... 50
Room with bath—resident..................... 60
Two rooms with bath.......................... 80
Three rooms with bath........................ 100
Public shower................................ 200
Public basins................................ 150
Slop sink.................................... 30
Office buildings:
White-collar worker (per person)*................ 2–3
Other workers (per person)................... 4.0
Cleaning per 10,000 sq ft..................... 30.0
Hospitals:
Per bed..................................... 80–100

* The value for white-collar workers is for office occupancy only, not including allowance for employees lunch rooms, dining rooms, etc. Requirements for these areas should be calculated separately.
Reprinted by permission from "ASHRAE Handbook and Product Directory—Systems, ASHRAE, New York, 1973.

Table 1-64. Estimated Hot Water Demand Characteristics for Various Types of Buildings

Type of building	Hot water required per person	Max. hourly demand in relation to day's use	Duration of peak load, hr	Storage capacity in relation to day's use	Heating capacity in relation to day's use
Residences, apartments, hotels, etc.†·‡	20–40 gpd*	⅐	4	⅕	⅐
Office buildings.....................	2–3 gpd*	⅕	2	⅕	⅙
Factory buildings..................	5 gpd*	⅓	1	⅖	⅛

* At 140°F.

† Daily hot water requirements and demand characteristics vary with the type of hotel. The better class hotel has a relatively high daily consumption with a low peak load. The commercial hotel has a lower daily consumption but a high peak load.

‡The increasing use of dishwashers and laundry washing machines in residences and apartments requires additional allowances of 15 gal per dishwasher and 40 gal per laundry washer.

Reprinted by permission from "ASHRAE Handbook and Product Directory—Systems," ASHRAE, New York, 1973.

Hot Water for Kitchens. Although, in private dwellings, a water temperature of 140°F is reasonable for dishwashing, in public places sanitation regulations call for 180°F water. Most of the dishwashing machines now available on the market require 180°F water. The amount of 180°F water needed in restaurants per day may be determined according to the American Gas Association method outlined in the following paragraphs:

1. Multiply the number of meals per day by the number of dishes per meal (6 for low-price restaurants, 8 for medium-price restaurants, and 10 for high-price restaurants) to determine the total number of dishes per day.

2. Divide the total number of dishes per day by the average number of dishes per rack to find the number of racks per day.

3. Multiply the number of racks per day by the gallons of 180°F water (using 1.5 gal for single-tank machines and 0.75 gal for two-tank machines). This product will give the gallons of 180°F water per day for rinse sprays.

4. Multiply the number of meal periods per day (one, two, or three) by the dishwashing tank capacity in gallons, giving the gallons of 180°F water per day necessary to fill the tanks.

5. Add values from (3) and (4) to obtain the total number of gallons of 180°F water required per day.

For purposes other than dishwashing, a considerable amount of 140°F water is used. To find the daily 140°F water requirement in a restaurant, multiply the total number of meals served per day by the gallons of 140°F water per meal. Low-price restaurants on the average utilize 0.9 gal of 140°F water per meal; medium- and high-price restaurants use 1.2 and 1.5 gal per meal, respectively.

Table 1-65. Hot Water Demand per Fixtures for Various Types of Buildings

Gallons of water per hour per fixture, calculated at a final temperature of 140 F

	Apartment House	Club	Gym-nasium	Hospital	Hotel	Industrial Plant	Office Building	Private Residence	School	Y.M.C.A.
1. Basins, private lavatory	2	2	2	2	2	2	2	2	2	2
2. Basins, public lavatory	4	6	8	6	8	12	6	....	15	8
3. Bathtubs	20	20	30	20	20	....	....	20	....	30
4. Dishwashers[a]	15	50–150	....	50–150	50–200	20–100	....	15	20–100	20–100
5. Foot basins	3	3	12	3	3	12	....	3	3	12
6. Kitchen sink	10	20	....	20	30	20	20	10	20	20
7. Laundry, stationary tubs	20	28	....	28	28	....	....	20	....	28
8. Pantry sink	5	10	....	10	10	....	10	5	10	10
9. Showers	30	150	225	75	75	225	30	30	225	225
10. Slop sink	20	20	....	20	30	20	20	15	20	20
11. Hydro-therapeutic showers				400						
12. Hubbard baths				600						
13. Leg baths				100						
14. Arm baths				35						
15. Sitz baths				30						
16. Continuous-flow baths				165						
17. Circular wash sinks				20	20	30	20		30	
18. Semi-circular wash sinks				10	10	15	10		15	
19. Demand factor[b]	0.30	0.30	0.40	0.25	0.25	0.40	0.30	0.30	0.40	0.40
20. Storage capacity factor[b]	1.25	0.90	1.00	0.60	0.80	1.00	2.00	0.70	1.00	1.00

[a] Dishwasher requirements should be taken from Table 13 or from manufacturers' data for the model to be used, if this is known.
[b] Ratio of storage tank capacity to probable maximum demand per hour. Storage capacity may be reduced where an unlimited supply of steam is available from a central street steam system or large boiler plant.

Reprinted by permission from "ASHRAE Handbook & Product Directory," ASHRAE, New York, 1973.

1-13. Gas Piping

Table 1-66. Capacity of Gas Piping, ft³/hr

At pressure drop of 0.3 in. water.   Specific gravity = 0.60

Pipe length, ft	Iron pipe size (IPS), in.								
	½	¾	1	1¼	1½	2	2½	3	4
10	132	278	520	1,050	1,600	3,050	4,800	8,500	17,500
20	92	190	350	730	1,100	2,100	3,300	5,900	12,000
30	73	152	285	590	890	1,650	2,700	4,700	9,700
40	63	130	245	500	760	1,450	2,300	4,100	8,300
50	56	115	215	440	670	1,270	2,000	3,600	7,400
60	50	105	195	400	610	1,150	1,850	3,250	6,800
70	46	96	180	370	560	1,050	1,700	3,000	6,200
80	43	90	170	350	530	990	1,600	2,800	5,800
90	40	84	160	320	490	930	1,500	2,600	5,400
100	38	79	150	305	460	870	1,400	2,500	5,100
125	34	72	130	275	410	780	1,250	2,200	4,500
150	31	64	120	250	380	710	1,130	2,000	4,100
175	28	59	110	225	350	650	1,050	1,850	3,800
200	26	55	100	210	320	610	980	1,700	3,500

From ANSI *Standard Installation of Gas Appliances and Gas Piping*, ANSI Z21.30-1959.   Reprinted by permission from "ASHRAE Handbook of Fundamentals," ASHRAE, New York, 1972.

Table 1-67. Multipliers for Various Specific Gravities

Sp gr	Multiplier	Sp gr	Multiplier	Sp gr	Multiplier
0.35	1.31	0.75	0.895	1.40	0.655
.40	1.23	0.80	.867	1.50	.633
.45	1.16	0.85	.841	1.60	.612
.50	1.10	0.90	.817	1.70	.594
.55	1.04	1.00	.775	1.80	.577
.60	1.00	1.10	.740	1.90	.565
.65	0.962	1.20	.707	2.00	.547
.70	.926	1.30	.680	2.10	.535

Reprinted by permission from "ASHRAE Handbook of Fundamentals," ASHRAE, New York, 1972.

Table 1-68. Common Gas Appliances

Maximum Gas Consumption in Ft³/Hour

Appliance	Natural gas 1050 Btu/ft³	Mixed gas 800 Btu/ft³	Manufactured gas 550 Btu/ft³
Range, domestic, 4 top, 1 oven burners...............	60	80	115
Range, domestic, 6 top, 2 oven burners...............	100	135	200
Hot plate, domestic or laundry stove per burner......	8.5	11	16
Room heater, radiant type, single, domestic..........	2	2.5	4
Water heater, instantaneous, automatic, per 1 gpm capacity.....................................	36	47	68
Refrigerator.....................................	2.6	3.1	4.5

ELECTRICAL

1-14. Power Systems

The typical circuit arrangements of power systems found in buildings may be classified as follows: radial, secondary selective, secondary (spot) network, and primary selective.

The radial arrangement employs a single power source and one circuit to each load. An equipment failure will result in a power outage until difficulty is corrected. The high quality of modern distribution equipment provides the service reliability which justifies the use of the radial arrangement for a majority of applications.

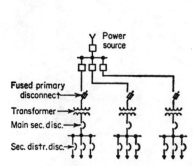
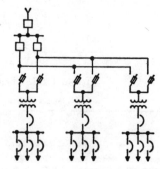

FIG. 1-21.   Radial-circuit arrangement.   FIG. 1-22.   Secondary-selective-circuit arrangement.

The secondary selective arrangement is in effect two radial systems with a secondary tie between them. It is provided in buildings where a greater degree of reliability is desired. This arrangement permits any secondary bus to be energized from either of two sources.

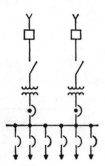
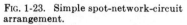

FIG. 1-23.   Simple spot-network-circuit arrangement.   FIG. 1-24.   Primary-selective-circuit arrangement.

The secondary-network arrangement is one where a high degree of service continuity is desired, as in large institutional buildings. The arrangement consists of two or more transformers energized by separate primary circuits, with the respective secondaries joined together.

The primary-selective arrangement provides an alternative power source to the substation transformers, but does not provide an alternative source of power to the secondary loads in event of a transformer outage.

The local prevailing rules of the Electric Service Company will usually determine the type and voltage of service available, regardless of the building size. This service may be from the secondary-network system in the street or for buildings of large magnitude, a spot network being instituted for the specific building load. The service voltage may be either 208Y/120 or 480Y/277 volts. In some of the current taller buildings, spot-network vaults are established by the Electric Service Company on intermediate floors in addition to the basement. In areas where buildings can be served at voltages greater than the utilization voltage, in the range of 2,400 to 13,800 volts, greater flexibility is available for circuit arrangements and in the selection and establishment of the utilization voltage.

1-15. System Voltages

Distribution systems may be classified according to voltages, levels used to carry the power directly to the branch circuits, or to load-center transformers or substations at which feeders to branch circuits originate.

The nominal system voltages listed in the left-hand columns of Table 1-69 are officially designated as standard nominal voltages in the United States by ANSI C84.1-1970. For the low-voltage systems the associated nominal system voltages in the right-hand column are obsolete and should not be used. For primary distribution voltage systems, the numbers in the right-hand column may designate an older system in which the voltage tolerance limits are maintained at a different level than the standard nominal system voltage.

Typical voltage-level utilization and application are outlined in Table 1-70.

Table 1-69. Nominal System Voltages

Standard nominal system voltages	Associated nominal system voltages
Low-voltage Systems	
120	110,115,125
120/240*	110/220, 115/230, 125/250
208Y/120*	216Y/125
240/120*	
240	230, 250
480Y/277*	416Y/240, 460Y/265
480*	440, 460
600	550, 575
Primary Distribution Voltage Systems	
2,400	2,200, 2,300
4,160Y/2,400	
4,160*	4,000
4,800	4,600
6,900	6,600, 7,200
8,320Y/4,800	
12,000Y/6,930	11,000, 11,500
12,470Y/7,200*	
13,200Y/7,620*	
13,800Y/7,970	
13,800*	14,400
20,780Y/12,000	
22,860Y/13,200	
23,000	
24,940Y/14,400*	
34,500Y/19,920*	
34,500	33,000

* Preferred standard nominal system voltages.
SOURCE: "IEEE Recommended Practice for Electric Power Systems in Commercial Buildings," IEEE Std. 241-1974, p. 56.

Table 1-70. Voltage Levels—Utilization and Application

Typical nominal voltage levels	Utilization	Application
120/240	Light and power (light at 120 volts, power at 120 and 240 volts)	Small loads such as individual homes, multifamily dwellings, and small commercial occupancies
208Y/120, 3 phase	Light and power (light at 120 volts, power at 120 volts and 208 volts, 1 phase, and 208 volts, 3 phase)	Commercial buildings and small industrial shops with limited electrical load
240	Power	Commercial and industrial buildings
480 600	Power	Commercial and industrial buildings with substantial motor loads
480Y/277, 3 phase	Light and power (light at 277 volts, 1 phase, and power at 480 volts, 3 phase)	Commercial and industrial buildings
2,400	Distribution	Industrial, heavy motor loads directly and lighting through transformation
4,163	Distribution	Large-area, spread-out commercial institutional buildings such as shopping centers, schools, and motels; supply load centers and transformers for lighting and power
4,800	Distribution	Industrial, with substations for stepping voltage to lower levels for lighting and power
12,470Y/7,200 . . . . 13,200Y/7,620 13,800	Distribution	Large industrial plans with substations for stepping voltage to lower levels for lighting and power

1-16. Building Loads

This subsection contains tables which permit the establishment of the anticipated electrical load for the building. With the area and the knowledge of the building utilization, the building-load density can be formulated. Total building load can be estimated by application of pertinent factors in Table 1-71. Individual building-load densities are obtained by application of pertinent items and factors in Tables 1-72 to 1-76. The demand load is obtained by application of items and factors contained in Table 1-77.

Table 1-71. Load Density in Representative Plants and Buildings

Type of industry	Light and power, volt-amp demand/ft²	Type of industry	Light and power, volt-amp demand/ft²
Commercial		Manufacturing	
Bank..	6–8	Appliance....................	7–12
Department store..............	8–11	Automotive...................	7.5–12
Hotel........................	6–9	Beet sugar refinery...........	19
Office building.................	6–14	Cigarette manufacture.........	11
Restaurant...................	12–18	Chemical.....................	10–15
Small store..................	5–8	Electronics, industrial.........	6–10
Shopping center...............	7–10	Foundry*.....................	11–15
School......................	4–7	Glass........................	1.5–8.5
		Heavy machinery.............	7–13
		Light machinery..............	11–15
		Metal fabricating and assembly	3–8
		Small device, industrial........	4.5–10
		Textile......................	12

* Large electric furnace loads are not included. They should be considered separately.
SOURCE: "Electrical Equipment Specifications Manual," Book III, Load Estimating Data Table 5.3, p. 2, General Electric Co., 1959.

Table 1-72. Approximate Electric Load, in Watts per Square Foot, for Various Footcandle Levels*

Foot-can-dle level	Coefficient of utilization																
	0.20	0.24	0.28	0.32	0.36	0.40	0.44	0.48	0.52	0.56	0.60	0.64	0.68	0.72	0.76	0.80
10	1.4	1.2	1.0	0.9	0.8	0.7	0.6	0.6	0.5	0.5	0.5	0.4	0.4	0.4	0.4	0.4
20	2.9	2.4	2.0	1.8	1.6	1.4	1.3	1.2	1.1	1.0	0.9	0.9	0.8	0.8	0.7	0.7
30	4.3	3.6	3.1	2.7	2.5	2.3	2.0	1.8	1.7	1.5	1.4	1.3	1.3	1.2	1.1	1.1
50	7.2	6.0	5.1	4.5	4.0	3.6	3.2	3.0	2.8	2.6	2.4	2.2	2.1	2.0	1.9	1.8
80	11.4	9.5	8.2	7.2	6.4	5.7	5.2	4.8	4.4	4.1	3.8	3.6	3.4	3.2	3.0	2.9
100	14.3	11.9	10.2	8.9	8.0	7.2	6.5	6.0	5.5	5.1	4.8	4.5	4.2	4.0	3 8	3.6

NOTE: Apply correction factor of Table 1-65 for specific light source.
* 50 lm/W; 0.70 in light-loss factor.
SOURCE: "IEEE Recommended Practice for Electric Power Systems in Commercial Buildings," IEEE Std. 241-1974, p. 286.

Table 1-73. Approximate Correction Factors for Some Typical Lamps

Lamp type	Cool white and warm white	Deluxe cool white and warm white
Fluorescent lamps:		
40 W T12, 430 mA	0.72	1.02
8-ft slimline, 430 mA	0.70	1.02
8-ft high output, 800 mA	0.67	0.95
8-ft extra high output, 1.5 A	0.71–0.78	1.02
Incandescent lamps:		
100 W		2.86
150 W		2.74
200 W		2.60
300 W		2.50
500 W		2.33
750 W		2.25
1000 W		2.14
Mercury lamps, deluxe white:		
100 W		1.39
175 W		1.20
250 W		1.14
400 W		0.97
700 W		0.89
1000 W		0.85
Metal halide lamps:		
400 W		0.64
High-pressure sodium lamps:		
400 W		0.48

NOTE: To find the correction factor for lamps not listed here, divide 50 by the lumens per watt of the lamp (including ballast losses). Lamp lumen data can be found in the catalogs of lamp manufacturers.
SOURCE: "IEEE Recommended Practice for Electric Power Systems in Commercial Buildings," IEEE Std. 241-1974, p. 286.

Table 1-74. Air-conditioning Load Density

Based on 1.5 Kva/Ton, Air-cooled Units*

Application	Demand	
Banks	4.5–6	va/ft²
Barber shops	5–6	va/ft²
Bars and taverns	165–210	va/seat
Beauty parlors	750	va/booth
Department stores:		
Main floor	7.5–10	va/ft²
Upper floors	4.5–6	va/ft²
Top floors	5–7.5	va/ft²
Bargain basement	6–10	va/ft²
Normal basement	4.5–6	va/ft²
Dress shops	4.5–10	va/ft²
Drugstores	4.5–10	va/ft²
Funeral parlors	3.75–5	va/ft²
Grocery stores	3.75–5	va/ft²
Night clubs:		
Convention type	190–210	va/seat
Week-end peak	165–190	va/seat
Offices:		
Multistory	3–3.75	va/ft²
Single floor	3.75–4.5	va/ft²
Top floor	5–6	va/ft²
Restaurants:		
Cafeterias	165–210	va/seat
Hotel dining-rooms	130	va/seat
Family restaurants	125–150	va/seat
Shoe shops	4.5–9.0	va/ft²
Supermarkets	3.75–5.25	va/ft²
Theaters:		
Continuous performances	82.5–100	va/seat
Neighborhood	75–82.5	va/seat

* For water-cooled units multiply load by 0.75.

Table 1-75. Air-conditioning Equipment

Kva Demand (Air-cooled)

Type of drive	Tons	Btu	Equipment kva demand
Induction-motor drive	1	12,000	1.5
0.8 PF synchronous-motor drive	1	12,000	1.65
1.0 PF synchronous-motor drive	1	12,000	1.3

SOURCE: "Electrical Equipment Specifications Manual," Book III, Load Estimating Data Tables 5.6 and 5.7, p. 3, General Electric Co., 1959.

Table 1-76. Kva-demand Material-handling Loads

Load	Kva demand*	Load	Kva demand*
Conveyors...............	1–15	Escalators..............	10–40
Cranes:		Hoists:	
Gantry.................	25–200	Ash and cinder........	1–5
Traveling bridge.......	5–200	Tramrail 1-ton........	1.5–3
Dumbwaiters............	1/2–5	Tramrail 5-ton........	6–10
Elevators:		Warehouse loading....	1–3
1-ton freight...........	3–20		
5-ton freight...........	7.5–20		
10-passenger...........	7.5–30		
20-passenger...........	7.5–50		
27-passenger...........	10–60		

* Demand depends upon rate of travel as well as size of load.
SOURCE: "Electrical Equipment Specifications Manual," Book III, Load Estimating Data Table 5.4, p. 2, General Electric Co., 1959.

Table 1-77. Demand Factors of Utilization Equipment

Equipment	Range, per unit	Equipment	Range, per unit
Arc furnaces.....................	0.90–0.100	Hand tools..................	0.20–0.40
Arc welders.....................	.20–.50	Induction furnaces and heating	
Compressors....................	.20–.50	equipment...................	.80–1.0
Conveyors.....................	.90–.100	Lighting......................	.75–1.0
Cranes.......................		Paper mills...................	.50–.70
		Resistance ovens, heaters,	
Elevators (quantity)		furnaces.....................	.80–1.0
		Resistance welders..............	.05–.40
1–2	1.0	Rubber mills..................	.50–.70
3	0.9	Pumps.......................	.20–.50
4	.775	Rolling mills..................	.20–.50
6	.55	Refineries....................	.50–.70
10	.48	Textile mills..................	.70–1.0
20	.44	Ventilation, blower motors.......	.20–.50

SOURCE: "Electrical Equipment Specifications Manual," Book III, Load Estimating Data Table 5.5, p. 2, General Electric Co., 1959.

1-17. Distribution

The purpose of any electric system is to provide a continuous supply of energy to the utilization equipment at reasonable cost. A typical power distribution system consisting of transformer, switchboard, motor control center, panelboards, feeders, lighting and power arrangements, and their relationship in an integrated system is outlined in Fig. 1-25.

Wire and cables in conduit and busways are used as feeders and capable of carrying large and small blocks of power from main switchboard to load centers to loads.

The feeder-conductor volt-drop limitations to building loads are outlined in Table 1-78.

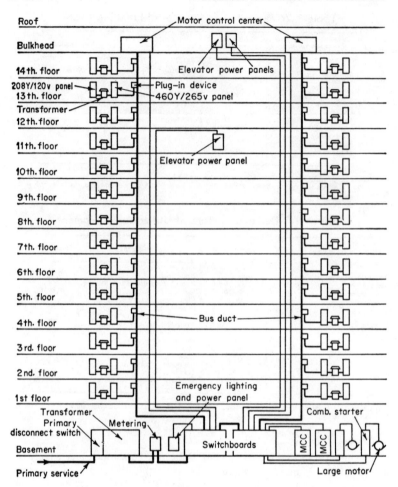

Fig. 1-25. Typical power-riser diagram.

Table 1-78. Feeder-conductor Volt-drop Limitations

Load	Limitations,* %
Power, heating or lighting or combination thereof...................	3
Max. total drop for conductors for feeders and branch circuits......	5

* Recommended by National Electrical Code 1975.

The IPCEA has published ampere ratings of cables insulated with oil-impregnated paper, varnished cambric, and rubber compounds.   The National

Electrical Code publishes current ratings of low-voltage cables for most applications in commercial buildings. Wiring in insured buildings must be installed at rated values which do not exceed those in the NEC or other local codes which are more restrictive than IPCEA ratings.

Because of the versatility, flexibility, and economic feasibility for the method of electrical distribution in large commercial and institutional buildings, feeder and plug-in busways are being widely accepted. Various bus ducts are listed in Table 1-79.

Table 1-79. Bus Ducts

Type	Volt rating	Ampere rating	Conductor
Plug-in	600 ac	100	Copper
Plug-in	600 ac-dc	225, 400, 600, 800, 1000	Copper or aluminum
Plug-in or feeder, low impedance	600 ac-dc	600, 800, 1000, 1350, 1600, 2000, 2500, 3000, 4000, 5000	Copper or aluminum
Plug-in or feeder, high frequency, 120 to 10,000 cycles/sec	800 ac	400, 500, 700	Copper or aluminum
Feeder	600 dc	225, 400, 600, 800, 1000, 1350, 1600, 2000, 2500, 3000, 4000, 5000	Copper or aluminum
Feeder, current limiting	600 ac	1000, 1350, 1600, 2000, 2500, 3000, 4000	Copper

Circuit breakers, fuses, safety switches, and combinations of these devices provide protection in a building distribution system against short circuits, overloads, and undervoltage by controlling the flow of current up to their respective rating. These devices may be contained in switchboards, panelboards, control centers, or in individual enclosures. Typical devices are listed in Table 1-80.

Table 1-80. Ampere Rating of Standard Fuses. Safety Switches, Pressure Switches, and Circuit Breakers (600 Volts or Less)

Single-element Fuses*

15	70	225	1,000
20	80	250	1,200
25	90	300	1,600
30	100	350	2,000
35	110	400	2,500
40	125	450	3,000
45	150	500	4,000
50	175	600	5,000
60	200	800	6,000

Enclosed General-purpose Safety Switches

30	100	400	800
60	200	600	1,200

Enclosed Pressure Switches

800	1,600	2,500	4,000
1,200	2,000	3,000	5,000

Nonadjustable Trip Circuit Breakers

15	70	200	400
20	100	225	500
30	125	250	600
40	150	300	700
50	175	350	800

*Dual-element fuses that provide both motor-running protection and short-circuit protection are available in a much greater range of sizes.

Table 1-81 outlines a basis for selection of panelboards.

Table 1-81. Panelboards

Usage	Type	Max. circuits per panel	Remarks
Lighting........	Switch and fuse or circuit breaker	42*	Includes lighting and appliance panelboards†
Power..........	Switch and fuse or circuit breaker	None	Physical size is limiting factor

* Where more than 42 circuits originate at one location, use two panels. Not more than 42 overcurrent devices shall be installed in a lighting and/or lighting and appliance panelboard, or cabinet.
† Lighting and appliance panelboard is defined as having more than 10 per cent of its overcurrent devices rated 30 amp or less, for which neutral connections are provided.

1-18. Motors and Controls

The general requirements for motor provisions are outlined in the National Electrical Code (NEC). The design of motor installations must conform to the code requirements and should include considerations for adequacy, flexibility, voltage drop, and safety.

Motors can be classified as outlined in Table 1-83.

The full-load currents (amperes) of representative motors at typical voltages are outlined in Tables 1-82 and 1-84. These data are useful in determining the wiring and setting of protective devices.

Table 1-82. Full-load Currents in Amperes Single-phase
Alternating-current Motors

Horsepower	115V	230V
⅙	4.4	2.2
¼	5.8	2.9
⅓	7.2	3.6
½	9.8	4.9
¾	13.8	6.9
1	16	8
1½	20	10
2	24	12
3	34	17
5	56	28
7½	80	40
10	100	50

The values of full-load currents are for motors running at usual speeds and motors with normal torque characteristics. Motors built for especially low speeds or high torques may have higher full-load currents, and multispeed motors will have full load current varying with speed, in which case the nameplate current rating shall be used.

To obtain full-load currents of 208- and 200-volt motors, increase corresponding 230-volt motor full-load currents by 10 and 15 percent, respectively.

The voltages listed are rated motor voltages. Corresponding nominal system voltages are 110 to 120 and 220 to 240.

Reproduced by permission from the 1975 National Electrical Code, copyright National Fire Protection Association, Boston, Mass.

Table 1-83. Classification of Motors

	Type	Speed characteristics	Full voltage			Hp range	Application—see footnotes (a) to (e)
			Starting torque	Starting current			
			Constant-speed Drive				
Polyphase a-c	Squirrel-cage general-purpose Design A	Constant	Normal 1–2.5 times	High 6–8 times		All	(a) Fans and (c) centrifugal pumps and centrifugal compressors
	Squirrel-cage Design B	Constant	Normal 1–2.5 times	Normal 5–6 times		Medium small	(a) Fans and centrifugal pumps and centrifugal compressors
	Squirrel-cage Design C	Constant	High 2–2.5 times	Normal 5–6 times		Medium small	(b) Reciprocating pumps and compressors (e) started loaded
	Squirrel-cage Design F	Constant	Low 1.25	Low 4 times		Medium large	Fans, centrifugal pumps, and compressors
	Wound rotor	Constant or variable	High 1–2.5 times (with secondary control)	Low 1–3 times (with secondary control)		All	(a) Hoists (b) reciprocating pumps and compressors (c) and frequent (e) or hard start
	Synchronous high speed	Exactly constant	Normal 0.75–1.75 times	Normal 5–7 times		Medium large	(a) Fans and centrifugal pumps and centrifugal compressors
	Synchronous low speed	Exactly constant	Low 0.3–0.4 times	Low 3–4 times		Medium large	(a) Reciprocating compressors starting unloaded
	Two-value capacitor	Constant	High	Normal		Small	(b) Pumps and compressors
	Permanent split capacitor	Constant	Low	Normal		Fractional	(a) Fans, blowers

	Motor	Speed			Size	Application
Single-phase a-c	Capacitor start	Constant	Moderate	Normal	Small fractional	(a) Fans and pumps
	Repulsion induction	Constant	High	Normal	Medium small	(a) Fans (b) pumps and compressors
	Split phase	Constant and adjustable	Normal	Normal	Fractional	(a) Fans (b) pumps and compressors (d) fans—direct

Adjustable-speed Drive

	Motor	Speed			Size	Application
Polyphase a-c	Squirrel-cage high slip; transformer adjustment	Variable	Normal	Normal	Medium small	(a) Fans
	Squirrel-cage separate winding or regrouped poles	Constant multispeed	Normal or high	Normal or low	All	(a) Fans, (b) pumps, and compressors (c) Fans
	Wound rotor	Variable	High (with secondary control)	Low (with secondary control)	All	(b) centrifugal pumps and compressors
Single-phase a-c	Repulsion	Variable	High	Normal	Low and fractional	(a) Fans, centrifugal pumps (b) compressors (d) Fans, direct
	Capacitor low-torque tapped winding	Variable two-speed	Low	Normal	Fractional	(d) Fans
	Capacitor low-torque transformer adjustment	Variable	Low	Low	Fractional	(d) Fans
	Split-phase regrouped poles	Constant	Normal	Normal	Fractional	(d) Fans

a Drives having medium or low starting torque and inertia WR^2 such as fans and centrifugal pumps or reciprocating pumps and compressors started unloaded.
b Drives having high starting torques, such as reciprocating pumps and compressors started loaded.
c Similar to (a) except where frequent or hard starting (large WR^2) requires a higher starting and accelerating torque.
d Fans direct-connected.
e Stoker drives.
f Torque depends on hp rating and speed. See *NEMA Standard* MG1-4.10, Motors and Generators.
SOURCE: "Heating, Ventilating, Air Conditioning Guide," vol. 37, pp. 642 and 643, ASHRAE, New York, 1959.

Table 1-84. Full-load-current* Three-phase A-C Motors

Hp	Induction-type squirrel-cage and wound-rotor, A					Synchronous-type unity power-factor,† A			
	115V	230V	460V	575V	2,300V	220V	440V	550V	2,300V
½	4	2	1	0.3					
¾	5.6	2.8	1.4	1.1					
1	7.2	3.6	1.8	1.4					
1½	10.4	5.2	2.6	2.1					
2	13.6	6.8	3.4	2.7					
3	....	9.6	4.8	3.9					
5	....	15.2	7.6	6.1					
7½	....	22	11	9					
10	....	28	14	11					
15	....	42	21	17					
20	....	54	27	22					
25	....	68	34	27	...	54	27	22	
30	....	80	40	32	...	65	33	26	
40	....	104	52	41	...	86	43	35	
50	....	130	65	52	...	108	54	44	
60	....	154	77	62	16	128	64	51	12
75	....	192	96	77	20	161	81	65	15
100	....	248	124	99	26	211	106	85	20
125	....	312	156	125	31	264	132	106	25
150	....	360	180	144	37	...	158	127	30
200	....	480	240	192	49	...	210	168	40

For full-load currents of 208- and 200-volt motors, increase the corresponding 230-volt motor full-load current by 10 and 15 per cent, respectively.

* These values of full-load current are for motors running at speeds usual for belted motors and motors with normal torque characteristics. Motors built for especially low speeds or high torques may require more running current, and multispeed motors will have full-load current varying with speed, in which case the nameplate current rating shall be used.

† For 90 and 80 per cent power factor the above figures shall be multiplied by 1.1 and 1.25 respectively. The voltages listed are rated motor voltages. Corresponding nominal system voltages are 110 to 120, 220 to 240, 440 to 480 and 550 to 600 volts.

Reproduced by permission from the 1975 National Electrical Code, copyright National Fire Protection Association, Boston, Mass.

1-19. Telephones

Well-planned communication facilities for both present and future needs incorporated into buildings during initial construction or major alterations will be beneficial throughout the life of the building.

Modern buildings may require teletypewriter service, data-transmission service—connections between data-processing machines over telephone facilities, centrex service—permitting the dialing of outgoing calls as well as receiving incoming calls without attendant, and public telephone service.

Communication needs of the building include:

Raceway. Underfloor ducts or cellular floor systems to serve as a telephone cable distribution facility.

Apparatus Closets. To house relay cabinets and auxiliary apparatus of modern key telephone systems.

PBX Equipment Rooms. To house the large equipment required for PBX service.

Cable Riser Systems. For bringing cables from the main terminal room to the various building floors.

Main Terminal Room. Connecting point between building and outside facilities.

Service from the Street. Aerial or underground service into the building.

A desirable distribution arrangement for commercial buildings is to divide the floor space into zones of no more than 10,000 ft², and preferably from 4,000 to 6,000 ft², to handle the distribution cables. The size of the raceway in each zone should be one square inch for every 100 ft² of office area, which is predicated upon the average allocation of one desk and telephone per 100 ft² of floor area. The raceway for the distribution cables can be provided as follows:

Underfloor Ducts. Spaced 4½ to 6 ft between parallel runs or feeds with cross runs and junction boxes located every 40 ft or less.

Cellular Floor. Appropriate cell utilization with header ducts connected to cell area at intervals for maximum coverage, usually no greater than 50 ft.

The communications equipment in each zone is connected to relay cabinets and other apparatus in a central closet in each zone. The type and size closet is outlined in Table 1-85.

Table 1-85. Telephone Zone Closets

Specification	Walk-in closet	Shallow closet
Depth:		
Minimum....................	3 ft	1½ ft
Maximum...................	None	2½ ft
Width:		
Minimum....................	5 ft	3 ft
Maximum...................	None	None
Floor area per 100 ft² served......	4 ft²	None
Length of walls per 1,000 ft² served	2½ ft	2½ ft
Minimum height of doors........	6 ft 8 in.	6 ft 8 in.*
Minimum width of doors.........	3 ft	3 ft†

* When shallow closets are used, the center post between double doors should be eliminated, if possible.
† Minimum for single door, 2½ ft for double doors.
SOURCE: Bell Telephone System, Telephone-planning Fact File, AIA File 31-i-5.

1-20. Signal and Communications Systems

Different types of buildings require a variety of signal and communications systems.

Industrial Buildings. *Burglar Alarm.* Burglar alarm system may be used to protect all doors, windows, elevator openings, skylights, etc.

Clock and Program Systems. These are used for indicating the time of day and operating signal devices such as bells or horns at predetermined times, such as starting and stopping work, rest periods, lunch periods, etc.

Door Alarm. Door alarm system is used to signal the guard room when certain restricted areas have been entered or vacated by individuals.

Fire Alarm. Fire alarm systems should be of the closed-circuit supervised type. Generally, noncoded systems are limited to small plants, since they only transmit a general alarm and do not indicate the location of the operated station. Coded systems are preferable.

Fire Detection. Automatic fire detection system may be used separately or combined with the manual fire alarm systems.

Intercom. Intercommunicating system may be provided in various forms.

P.A. Public address sound system may be used throughout the plant for paging, radio programs, recordings, announcements, and entertainment.

Paging. Paging system is used to call and locate individuals.

Smoke Detection. This system is used to detect smoke in ventilating, air-conditioning, and dust-collecting ducts.

Sprinkler System. This alarm system is used to signal when sprinkler heads open, when noticeable leaks occur, when water flow valves operate in either dry or wet systems, when post indicator valves operate or are left open, or when the shutoff valves are placed in any subnormal position.

Watchman's System. Watchman's supervisory system should be of a type which will require the watchmen on the various tours to produce a record in the superintendent's or chief guard's quarters at the start and at the finish of each tour.

Commercial Buildings. *Fire Systems.* Fire alarm system of the manual type should be provided for the protection of the general public and the employees within the building.

A fire alarm system of the automatic type should be provided where records or files are kept or stored. A fire-line signal system is used exclusively by members of the fire department to transmit signals to the pump room.

A fire-line telephone system consists of a master station telephone located in the pump room, submaster station telephones located in the auxiliary pump room and at the building entrance, and outlying telephones located on each of the other floors. This system is of the common-talking type.

Schools. *Clock and Program System.* Clock and program system provides the means of showing correct time throughout the premises and to denote the different periods in a day's schedule.

Fire Alarms. Any fire signal should be distinctive from all other signals and should be audible to everyone in the building.

A fire alarm system for use in schools is one of four types, having a common characteristic: they are all closed-circuit, electrically supervised.

In small schools, either the noncoded or master-coded type is frequently used.

In large schools and colleges, the coded types of system are used.

Intercom System. Telephones are used for intercommunication between the principal's office and the main office and the classrooms.

Sound and Radio Distribution System. A sound and radio distribution system enables the distribution of radio programs, recordings, lectures, and announcements.

Hospitals. *Clock System.* A clock system is important in hospitals, both for keeping time and for administering anesthesia.

Emergency Call. An emergency feature may be added to any nurse-call system. This is used by the nurse to call assistance to a patient's room when the occasion requires.

Fire Alarm System. A fire alarm system for use in hospitals is usually of the General Alarm type.

Nurse Call. A nurse-call system is used by patients to call a nurse to the bedside. There are two general types of such systems, the visual and the audio.

Paging Systems. Paging systems are used to locate doctors and other members of the staff throughout the building. The visual system uses lamp annunciators throughout and also incorporates an audible signal such as a buzzer or chime.

The sound system consists of loudspeakers throughout one or more hospital buildings.

"In" and "out" systems are used by the doctors and other members of the staff to designate whether or not they are in the building.

Sound and Radio System. Sound and radio systems enable patients to listen in on one or more channels of radio programs, TV, recordings, announcements, etc.

1-21. Grounding

The purpose of grounding is to provide protection of personnel, equipment, and circuits by eliminating the possibility of dangerous or excessive voltage. The considerations of grounding may be subdivided as follows:

System Grounding. In an electric power distribution system system grounding is concerned with the nature and location of an intentional electric interconnection between the electric system conductors and ground (earth). Specific methods of power system grounding include the underground system, solid neutral grounding, low-resistance grounding, high-resistance grounding, high-resistance grounding with traceable signal at fault, corner-of-the-delta grounding, mid-phase grounding and low-reactance grounding.

From a practical standpoint, two methods of low-voltage system grounding, namely the solidly grounded neutral and high-resistance grounded neutral methods, will best fulfill the majority of application requirements. Principal characteristics of major methods of grounding low-voltage systems are outlined in Table 1-86.

Equipment Grounding. This designation pertains to that system of electric conductors by which all metallic structures, through which energized conductors run, will be interconnected. The equipment grounding ensures freedom from electric shock to personnel by maintaining a low potential difference between nearby metallic members and provides an adequate and effective electric conductor over which short-circuit currents involving ground can flow without sparking or other thermal distress so as to avoid a fire hazard to combustible material.

Static Grounding. Static grounding is concerned with the connection to ground of static accumulation on equipment, materials being handled, or even on operating personnel in a manner to eliminate a potentially hazardous operating condition, where the discharge of a static charge to ground or other equipment in the presence of flammable or explosive materials can cause fires and explosions.

Lightning Protection Gounding. The conduction to earth of current discharges in the atmosphere originating from electric charges in cloud formations is lightning protection grounding. The function of the lightning grounding system is to convey the lightning discharge currents safely to earth without incurring damaging potential differences across electrical insulation in the power system, without overheating the lighting grounding conductors, and without the disrupting breakdown of air between the lightning rod conductors and other metallic members of the structure.

Connection to Earth. This is one of the most important parts of the grounding system and should be one of low resistance. For large substations, the earth resistance should not exceed 1 ohm. For smaller substations and industrial plants, an earth resistance of less than 5 ohms should be obtained if practicable. The National Electrical Code (1975) states that the maximum resistance shall not exceed 25 ohms. Ground electrodes may be underground metallic piping systems, metal building, framework, well casings, steel piling, and other underground metal structures installed for purposes other than grounding, or they may be made electrodes specifically designed for grounding purposes. Made electrodes may be driven rods, buried strips or cables, grids, buried plates, and counterpoises.

Table 1-86. Principal Characteristics of Major Methods of Grounding Low-voltage Systems

System property	Type of system grounding		
	Solid*	High resistance†	Ungrounded
Immediate shutdown of faulty circuit on occurrence of first ground fault	Yes	No	No
Control of transient overvoltages due to arcing ground faults	Yes	Yes	No
Control of impressed or self-generated steady-state overvoltages	Yes	No	No
Flash hazard to personnel during ground fault (no escalation of fault)	Severe	Essentially zero	Essentially zero
Arcing fault damage to equipment during ground fault (no escalation)	May be severe unless fault is promptly removed	Usually minor unless fault removal is so prolonged as to cause fault escalation	Usually minor but transient over-voltages may cause fault escalation or multiple insulation failures
Shock hazard, unfaulted phases to ground, during ground fault	Line-to-neutral voltage	Approximately line-to-line voltage	May be several times line-to-neutral voltage
Shock hazard, equipment frame to ground during solid internal line-to-ground fault	Moderate	Minimum	Small
Detection of arcing faults	L-L or L-G arcing faults readily detected, esp. with ground fault relaying	Ground detectors and fault locating equipment required for L-G arcing faults. L-L faults readily detected by phase overcurrent devices unless fault current is severely limited	Ground detectors and fault locating equipment required for L-G arcing faults. Transient over-voltages may meanwhile cause additional insulation breakdowns. L-L faults readily detected by phase overcurrent devices unless fault current is severely limited
Suitable for four-wire, three-phase service	Yes	No	No

* For optimum results, use of solid grounding method should include sensitive ground fault relaying.
† For optimum results, use of high-resistance grounding method should include equipment and procedures for alarming, tracing, and removing the ground fault promptly.
SOURCE: General Electric Electrical Equipment Specifiers' Guide, 1972 11-AP-3, p. 37.

SECTION 2

ILLUMINATION

D. L. Whitehead, M.S.; Manager, Engineering Laboratories, High Voltage Section, Westinghouse Electric Corporation; Fellow, Institute of Electrical and Electronics Engineers; Committee Member, National Electrical Manufacturers Association, ASA

CONTENTS

Definition of Terms   2–2

Inverse-square Law   2–2

Application   2–2

ILLUMINATION

Definition of Terms

Brightness (*B*) is that property of a light source that specifies the ability
of an element of the source to produce luminous effects. It may be expressed
in two ways: candles per unit area, such as candles/sq in. (c/in.²), or lumens
per unit area.

Footcandle (fc) is the illumination at a point on a surface which is one foot
from and perpendicular to a uniform point source of one candle.

Footlambert (fl) is the brightness of a surface emitting or reflecting one lumen
per square foot.

Footlambert (fl) = footcandles (fc) × reflection factor

$$= \frac{\text{lumens (incident)} \times \text{reflection factor}}{\text{area (sq ft) of surface}} \qquad (2\text{-}1)$$

Illumination (*E*) is the density of luminous flux on a given surface. The
unit of measure is the footcandle.

Lambert (*B'*) is the brightness of a surface emitting or reflecting one lumen
per square centimeter.

Lumen (lm) is the quantity of light flux falling on a surface of one square
foot from a uniform point source of one candle. A one-square-foot section
from a sphere of one-foot radius with a one-candle source at its center would
be such a surface. The lumen differs from the candle in that it is a measure
of light flux, irrespective of direction.

$$\text{Incident lumens} = \text{footcandles} \times \text{area (sq ft)} \qquad (2\text{-}2)$$

Luminous flux (*F*) is the time rate of flow of light. The unit of measure
is the lumen.

Luminous intensity (*I*) (candlepower) is that property of a light source
which specifies its ability as a whole to produce luminous effects. The standard unit of intensity in a given direction is the International Candle. An
ordinary wax candle has a luminous intensity of approximately one candle.

$$\text{Candlepower (cp)} = \text{footcandles (fc)} \times \text{distance squared } (D^2) \qquad (2\text{-}3)$$

where D = distance in feet from light source to illuminated surface.

Mean spherical candlepower (MSCP) is the average candlepower of a source
in all directions.

$$\text{MSCP} = \frac{\text{lumens}}{12.57} \qquad (2\text{-}4)$$

Inverse-square Law. Illumination decreases inversely as the square of the
distance. When the light rays are perpendicular to the surface:

$$\text{Illumination } E = \frac{\text{luminous intensity}}{\text{distance squared}} = \frac{I}{D^2} \qquad \text{footcandles} \qquad (2\text{-}5)$$

where I is the candlepower and D is the distance in feet.

When the light rays are not perpendicular to the surface, the horizontal illumination is

$$E_h = \frac{I \times \cos \theta}{D^2} \qquad \text{footcandles} \qquad (2\text{-}6)$$

and the vertical illumination is

$$E_v = \frac{I \times \sin \theta}{D^2} \qquad \text{footcandles} \qquad (2\text{-}7)$$

In both cases θ is measured from the vertical.

Table 2-1. Illumination Conversion Factors

Quantity	Multiply number of	By	To obtain
Brightness..........	blondels	0.0002054	Candles/sq in.
Brightness..........	Candles/sq cm	6.45	Candles/sq in.
Brightness..........	footlamberts	0.002210	Candles/sq in.
Brightness..........	lamberts	2.054	Candles/sq in.
Brightness..........	blondels	0.00003183	Stilbs
Brightness..........	Candles/sq cm	1	Stilbs
Brightness..........	Candles/sq in.	0.1550	Stilbs
Brightness..........	footlamberts	0.0003425	Stilbs
Brightness..........	lamberts	0.3183	Stilbs
Brightness..........	Candles/sq cm	31,416	blondels
Brightness..........	Candles/sq in.	4,870	blondels
Brightness..........	footlamberts	10.76	blondels
Brightness..........	lamberts	10,000	blondels
Brightness..........	Stilbs	31,416	blondels
Brightness..........	blondels	0.0929	footlamberts
Brightness..........	Candles/sq cm	2,919	footlamberts
Brightness..........	Candles/sq in.	452	footlamberts
Brightness..........	lamberts	929	footlamberts
Brightness..........	Stilbs	2,919	footlamberts
Brightness..........	blondels	0.0001	lamberts
Brightness..........	Candles/sq cm	3.1416	lamberts
Brightness..........	Candles/sq in.	0.487	lamberts
Brightness..........	footlamberts	0.001076	lamberts
Illumination........	lumens/sq cm	929	footcandles
Illumination........	lumens/sq meter	0.0929	footcandles
Illumination........	lumens/sq ft	1	footcandles
Illumination........	lux	0.0929	footcandles
Illumination........	phot	929	footcandles
Illumination........	footcandles	10.76	lux
Illumination........	lumens/sq cm	10,000	lux
Illumination........	lumens/sq meter	1	lux
Illumination........	lumens/sq ft	10.76	lux
Illumination........	phot	10,000	lux
Illumination........	foot-candles	0.001076	phot
Illumination........	lumens/sq cm	1	phot
Illumination........	lumens/sq meter	0.0001	phot
Illumination........	lumens/sq ft	0.001076	phot
Illumination........	lux	0.0001	phot
Luminous flux.......	light-watts	680	lumens
Luminous flux.......	youngs	680	lumens

Application. Generally acceptable lighting levels in terms of footcandles for various types of installations are listed in Table 2-2. This is the average illumination at the work level. The number of lamps required to produce a required level of illumination is given by

Number of lamps

$$= \frac{\text{footcandles} \times \text{area}}{\text{lumens per lamp} \times \text{coefficient of utilization} \times \text{maintenance factor}} \qquad (2\text{-}8)$$

The lumens per lamp for a number of standard bulbs of incandescent, mercury, and fluorescent types are listed in Tables 2-3 to 2-5. The coefficient of utilization and maintenance factor (MF) are selected from Table 2-6, corresponding to the luminaire that is to be used. A luminaire usually consists of a number of lamps. The total number of luminaires required is

$$\text{Number of luminaires} = \frac{\text{number of lamps}}{\text{lamps per luminaire}} \qquad (2\text{-}9)$$

To use Table 2-6, it is first necessary to determine a room index from Table 2-7, which covers a wide variety of room dimensions and light-mounting heights. Table 2-8 gives average data on diffuse-reflection ratios, and Table 2-9 lists maximum-brightness ratios that are acceptable.

Brightness ratios are determined as follows: Determine the reflection values of the task, desk, floor, walls, and ceilings from Table 2-8. The brightness of the task and desk is determined by multiplying the average foot-candles by the reflection factor. Brightness of walls, ceilings, and floor is determined from Table 2-11.

Table 2-2.   Illumination Levels, Interior Lighting

	Foot-candles Maintained in Service (*Not Initial Values*)
Assembly (manufacturing):	
Rough	20
Medium	50
Fine	100
Extra fine	300*
Auditoriums:	
Assembly only	10
Exhibitions	30
Banks:	
Lobby	20
Cages and offices	50
Barber shops and beauty parlors	50
Bathrooms:	
General lighting	5
At mirror (on face)	40
Bedrooms:	
General lighting	5
At mirror (on face)	20
Churches:	
Auditorium	10
Sunday-school rooms	20
Pulpit	20
Classrooms, on desks and chalkboards:	
Typical	30
Sight-saving or special	50
Depots and stations:	
Waiting room	20
Ticket rack and counter	50
Concourse	5
Platforms	5
Dining rooms:	
Homes (general lighting)	5
Hotels and restaurants	10
Drafting rooms	50
Elevators	10
Garages:	
Storage	10
Repair and servicing	50
Gymnasiums:	
Exhibitions and matches	30
General exercise	20
Assemblies	10
Dances	5
Lockers and shower rooms	10
Halls and corridors	5
Homes (see specific rooms)	
Hospitals:	
Private rooms and wards:	
General lighting	5
Supplementary for reading	20
Surgery:	
General lighting	50
Operating table	1800
Obstetrical:	
Delivery room	50
Delivery table	200
Examination table	50
Inspection:	
Rough	20
Medium	50
Fine	100
Extra fine	200 or more*
Ironing	40
Kitchens:	
General lighting	10
Supplementary (at task)	40

Table 2-2.   Illumination Levels, Interior Lighting (Continued)

	Foot-candles Maintained in Service (Not Initial Values)
Laboratories:	
General lighting	30
Work tables	50
Close work	100
Living rooms (see also specific visual task):	
General lighting	5
Lobbies	20
Machine shops:	
Rough bench and machine work	20
Medium bench and machine work	50
Fine bench and machine work	100
Extra fine bench and machine work	200 or more*
Mail rooms	30
Museums and art galleries:	
General lighting	10
On displays	50
Offices:	
Casual visual tasks: inactive file rooms, reception rooms, stairways, washrooms, and other service areas	10
Ordinary visual tasks: general office work (except for work classified as "difficult visual tasks"), private office work, general correspondence, conference rooms, active file rooms, mail rooms	30
Difficult visual tasks: auditing and accounting, business-machine operation, transcribing and tabulation, bookkeeping, drafting, designing	50
Reading:	
Short periods, material of reasonably good visibility	20
Prolonged periods or smaller type	40
Proofreading	100
Schools (see specific rooms)	
Sewing:	
Coarse work, high contrast between thread and fabric	20
Light fabrics, occasional periods	40
Light to medium fabrics, prolonged periods	80
Dark fabrics, fine detail, low contrast	150 or more*
Show windows:	
Low surrounding brightness:	
General displays	50
Feature displays	100
Medium surrounding brightness:	
General displays	100
Feature displays	200
High surrounding brightness:	
General displays	200
Feature displays	500
Stairways	10
Storage and stock rooms:	
Rough bulky material	5
Medium material	10
Fine material requiring care	20
Store interiors:	
Circulation areas	20
General merchandising areas	50
Showcases, wall cases, and open-counter displays	100*
Feature displays	200*
Theaters and motion-picture houses:	
Auditorium during intermission	5
Auditorium during picture	0.1
Foyer	5
Lobby	20
Toilets and washrooms	10
Waiting rooms	20
Woodworking:	
Rough sawing and bench work	30
Sizing, planing, rough sanding, veneering, medium machine and bench work	50
Fine bench and machine work, fine sanding and finishing	100
Writing	20

* Usually obtained by supplementary luminaires in combination with general lighting systems providing not less than one-tenth of the recommended value for the task.

SOURCE: "Foot Candle Tables," Bulletin A-4981, p. 3, Westinghouse Electric Corporation, Bloomfield, N.J.

Table 2-3. Incandescent-lamp Data

Watts	Bulb	Base	Finish	Rated avg life, hr	Initial lumens
			General-service Lamps		
100	A-21	Med.	I.f	750	1,620
150	A-23	Med.	I.f-cl.	750	2,600
200	PS-30	Med.	I.f.-cl.	750	3,700
300	PS-30	Med.	I.f.-cl.	750	5,900
300	PS-35	Mogul	I.f.-cl.	1,000	5,650
500	PS-40	Mogul	I.f.-cl.	1,000	9,900
750	PS-52	Mogul	I.f.-cl.	1,000	15,600
1,000	PS-52	Mogul	I.f.-cl.	1,000	21,500
1,500	PS-52	Mogul	I.f.-cl.	1,000	33,000
			Projector and Reflector Lamps		
75	PAR-38	Med. skt.	Projector spot	1,000	450 (0-15°)
75	PAR-38	Med. skt.	Projector flood	1,000	550 (0-30°)
150	PAR-38	Med. skt.	Projector spot	1,000	1,150 (0-15°)
150	PAR-38	Med. skt.	Projector flood	1,000	1,400 (0-30°)
75	R-30	Med.	Reflector spot	1,000	220 (0-15°)
75	R-30	Med.	Reflector flood	1,000	300 (0-30°)
150	R-40	Med.	Reflector spot	1,000	600 (0-15°)
150	R-40	Med.	Reflector flood	1,000	800 (0-30°)
300	R-40	Med.	Reflector spot	1,000	1,350 (0-15°)
300	R-40	Med.	Reflector flood	1,000	1,600 (0-30°)

SOURCE: "Illumination Design Data for Interiors," p. 6, Westinghouse Electric Corporation, Bloomfield, N.J.

Table 2-4. Mercury-lamp Data

Designation	Watts	Bulb	Base	Ballast loss/lamp, watts	Rated avg life, hr*	Initial lumens
A-H1	400	T-16	Mogul	40†	4,000	16,000
A-H12	1,000	T-28	Mogul	85†	3,000	60,000
A-H9	3,000	T-9½	S.C. term	165‡	5,000	120,000

* Rated average life under specified test conditions at 5 hr per start. At 10 hr/start, rated average life is 6,000 hr.
† Single lamp high PF 110 to 125-volt ballasts. Losses for two-lamp ballasts are generally lower.
‡ Single lamp high PF, 230-volt ballast.
SOURCE: "Illumination Design Data for Interiors," p. 6, Westinghouse Electric Corporation, Bloomfield, N.J

Table 2-5. Fluorescent-lamp Data

Bulb	Watts	Base	Rated avg life, hr	Rated initial lumens*		
				White	Std cool white	Std warm white
Preheat Lamps						
33″ T-12	25	Med. bipin	7,500†	1,430	1,370	1,440
48″ T-12	40	Med. bipin	7,500†	2,480	2,370	2,500
60″ T-17	90	Mog. bipin	7,500†	4,860	4,650	4,900
Instant-start Lamps						
48″ T-12	40	Med. bipin	6,000‡	2,480	2,370	2,500
60″ T-17	40	Mog. bipin	6,000‡	.....	2,300	
Slimline Lamps¶						
48″ T-12	38§	Single pin	6,000‡	2,320	2,200	2,340
	52			3,020	2,870	3,050
72″ T-12	59	Single pin	6,000‡	3,660	3,500	3,700
	72			4,300	4,100	4,340
96″ T-12	75	Single pin	6,000‡	4,800	4,575	4,850
	96			5,800	5,540	5,860
96″ T-8	34	Single pin	6,000‡	2,280	2,180	2,300
	51			3,300	3,150	3,330
	69			4,350	4,150	4,390

* Lumens measured after 100 hr burning at 80°F ambient and under specified test conditions.   The lumen outputs of the de luxe cool white and de luxe warm white lamps are approximately 40 per cent less than those of the corresponding standard cool white and standard warm white.   The lumen values of daylight and soft white lamps are 85 and 73 per cent, respectively, of the white values.

† Life under specified test conditions at 3 burning hours per start.   Lamp life is slightly longer for more burning hours per start.

‡ Life (tentative) under specified test conditions at 12 burning hours per start.   Lamp life is somewhat shorter for fewer burning hours per start.

¶ Slimlines may be operated at any current density within their design range.   The figures listed for the 96″ T-8 Slimline are for 120, 200, and 300 ma.   The data listed for the T-12 Slimlines are for 425 and 600 ma.

§ Operates on a standard 40-watt instant-start ballast at 420 ma.

source: "Illumination Design Data for Interiors," p. 6, Westinghouse Electric Corporation, Bloomfield, N.J.

Table 2-6. Coefficients of Utilization

For Explanation of Symbols, See Notes on Page 2-8

Luminaire	Ceiling..	75%			50%			30%	
	Walls...	50%	30%	10%	50%	30%	10%	30%	10%
	Room index	Coefficient of utilization							
MF G–.75↑0 M–.65 P–.55↓79 — Direct, RLM dome reflector, MS = 1.0 x MH	J	.37	.31	.27	.36	.31	.27	.31	.27
	I	.45	.41	.38	.45	.40	.37	.40	.37
	H	.49	.45	.42	.49	.45	.42	.45	.42
	G	.53	.49	.46	.53	.49	.46	.48	.46
	F	.56	.53	.49	.55	.52	.49	.51	.49
	E	.61	.58	.55	.60	.57	.55	.56	.55
	D	.66	.63	.60	.64	.62	.60	.61	.60
	C	.67	.65	.62	.66	.64	.62	.63	.61
	B	.71	.68	.66	.69	.67	.65	.66	.64
	A	.72	.70	.67	.71	.68	.67	.67	.66
MF G–.75↑0 M–.65 P–.55↓70 — Direct, RLM deep-bowl reflector, MS = 1.0 x MH	J	.35	.31	.28	.34	.31	.28	.30	.28
	I	.43	.39	.37	.42	.39	.37	.39	.37
	H	.46	.44	.42	.46	.44	.42	.43	.42
	G	.50	.47	.45	.49	.47	.45	.46	.45
	F	.53	.50	.47	.51	.49	.47	.49	.47
	E	.56	.54	.51	.56	.54	.51	.53	.51
	D	.61	.58	.56	.59	.57	.56	.56	.56
	C	.62	.60	.57	.61	.58	.57	.58	.56
	B	.64	.61	.61	.63	.61	.60	.60	.59
	A	.65	.63	.61	.64	.62	.61	.61	.60
MF G–.75↑0 M–.60 P–.40↓75 — Direct, high bay, narrow spread, MS =.6 x MH	J	.43	.40	.39	.42	.40	.39	.40	.38
	I	.51	.50	.49	.50	.49	.48	.49	.46
	H	.55	.54	.53	.54	.53	.52	.53	.52
	G	.59	.58	.57	.58	.56	.55	.56	.55
	F	.61	.60	.58	.59	.58	.58	.58	.57
	E	.64	.63	.62	.63	.62	.61	.61	.60
	D	.68	.65	.64	.66	.65	.64	.64	.63
	C	.69	.67	.66	.67	.66	.64	.64	.64
	B	.70	.68	.67	.68	.67	.66	.66	.65
	A	.71	.70	.68	.69	.67	.67	.67	.66
MF G–.75↑0 M–.65 P–.50↓75 — Direct, high bay, medium or wide spread, MS =1.0 x MH	J	.40	.36	.34	.39	.36	.34	.36	.33
	I	.48	.45	.43	.47	.44	.43	.44	.42
	H	.52	.50	.48	.51	.49	.47	.49	.47
	G	.55	.53	.52	.55	.52	.51	.52	.51
	F	.58	.56	.53	.56	.55	.53	.55	.53
	E	.62	.60	.56	.61	.59	.57	.58	.57
	D	.66	.63	.61	.64	.62	.61	.62	.61
	C	.67	.65	.62	.66	.64	–62	.63	.62
	B	.69	.67	.66	.67	.65	.64	.65	.64
	A	.70	.68	.67	.68	.67	.65	.66	.64
MF G–.80↑0 M–.72 P–.65↓70 — Direct, heavy duty narrow spread, MS =.5 x MH; medium spread, MS =.8 x MH	J	.40	.38	.36	.39	.38	.36	.38	.36
	I	.48	.46	.45	.47	.46	.45	.45	.43
	H	.52	.51	.50	.51	.50	.49	.50	.48
	G	.55	.54	.53	.54	.53	.52	.53	.51
	F	.57	.56	.55	.56	.55	.54	.55	.53
	E	.60	.59	.58	.59	.58	.57	.57	.56
	D	.64	.61	.60	.62	.60	.59	.60	.59
	C	.64	.63	.61	.63	.62	.60	.60	.60
	B	.65	.64	.63	.64	.63	.62	.62	.61
	A	.66	.65	.64	.64	.63	.62	.62	.62

Table 2-6. Coefficients of Utilization (Continued)

Luminaire	Ceiling..	75%			50%			30%	
	Walls...	50%	30%	10%	50%	30%	10%	30%	10%
	Room index	Coefficient of utilization							
MF G –.80 ↑0 M –.72 P –.65 ↓70 Direct, heavy duty, wide spread, MS = 1.1 x MH	J I H G F E D C B A	.37 .45 .48 .52 .55 .57 .62 .63 .64 .66	.34 .42 .46 .50 .52 .56 .59 .61 .62 .64	.31 .41 .45 .48 .51 .54 .57 .58 .61 .62	.36 .44 .49 .51 .50 .57 .60 .62 .63 .64	.34 .41 .45 .49 .51 .55 .58 .59 .61 .62	.31 .40 .44 .48 .50 .53 .57 .58 .60 .61	.34 .41 .45 .49 .51 .55 .57 .59 .60 .62	.31 .39 .44 .48 .50 .53 .57 .57 .59 .60
MF G –.70 ↑5 M –.60 P –.45 ↓58 Direct, RLM Glassteel diffuser, MS = 1.0 x MH	J I H G F E D C B A	.27 .34 .37 .40 .42 .46 .49 .51 .53 .54	.23 .30 .34 .37 .39 .43 .47 .49 .51 .53	.20 .28 .31 .34 .37 .41 .44 .46 .49 .51	.26 .33 .36 .39 .40 .45 .48 .49 .51 .53	.23 .29 .33 .36 .38 .42 .46 .47 .49 .51	.20 .27 .31 .34 .36 .40 .44 .46 .48 .49	.22 .29 .32 .35 .37 .41 .44 .46 .48 .49	.20 .27 .30 .33 .36 .40 .43 .44 .47 .48
MF G –.60 ↑0 M –.50 P –.40 ↓67 Direct, RLM silvered-bowl diffuser, MS =.8 x MH	J I H G F E D C B A	.38 .46 .49 .53 .55 .57 .61 .62 .63 .64	.36 .45 .49 .52 .54 .57 .59 .61 .62 .63	.35 .44 .48 .51 .53 .56 .58 .60 .61 .62	.38 .45 .49 .52 .53 .57 .59 .60 .61 .62	.36 .44 .48 .51 .53 .56 .58 .59 .60 .61	.35 .43 .47 .50 .52 .55 .57 .58 .59 .60	.36 .44 .48 .51 .53 .55 .57 .58 .59 .60	.35 .42 .47 .49 .51 .54 .56 .57 .58 .59
MF G –.75 ↑0 M –.65 P –.55 ↓65 Direct, vapor-tight, wide spread, MS = 1.0 x MH	J I H G F E D C B A	.31 .38 .41 .47 .51 .55 .56 .59 .60	.26 .34 .38 .44 .48 .52 .54 .57 .58	.23 .31 .34 .41 .46 .50 .52 .55 .56	.30 .37 .41 .46 .50 .54 .55 .58 .59	.26 .33 .38 .43 .48 .52 .53 .56 .57	.23 .31 .34 .41 .46 .50 .52 .54 .56	.26 .33 .37 .43 .47 .51 .52 .55 .56	.23 .31 .34 .41 .46 .50 .51 .54 .55
MF G –.70 ↑0 M –.60 P –.50 ↓53 Direct, prismatic lens, medium spread, MS =.8 x MH	J I H G F E D C B A	.25 .31 .34 .36 .38 .40 .43 .45 .48 .50	.22 .28 .31 .33 .35 .39 .41 .43 .45 .47	.20 .26 .29 .32 .34 .38 .40 .42 .44 .46	.24 .29 .32 .34 .36 .39 .42 .44 .47 .48	.22 .28 .31 .33 .34 .37 .40 .41 .43 .46	.20 .26 .29 .31 .33 .36 .39 .40 .42 .45	.22 .28 .30 .32 .34 .37 .39 .40 .42 .45	.20 .26 .28 .30 .32 .35 .38 .40 .41 .42

Table 2-6. Coefficients of Utilization (Continued)

Luminaire	Ceiling..	75%			50%			30%	
	Walls...	50%	30%	10%	50%	30%	10%	30%	10%
	Room index	Coefficient of utilization							

MF-.75 ↑0 ↓62
Direct, PAR-38, 150-watt shielded to 45°, total lamp lumens = 1850, MS=.5 x MH

Room index	50%	30%	10%	50%	30%	10%	30%	10%
J	.52	.49	.47	.51	.49	.47	.48	.47
I	.55	.53	.51	.54	.52	.51	.51	.50
H	.57	.55	.53	.56	.54	.53	.53	.53
G	.58	.57	.55	.57	.56	.55	.55	.54
F	.59	.58	.57	.58	.57	.56	.56	.56
E	.61	.60	.59	.60	.59	.58	.58	.57
D	.63	.62	.61	.61	.61	.60	.60	.59
C	.64	.64	.63	.63	.63	.62	.62	.61
B	.65	.65	.64	.64	.64	.63	.63	.62
A	.66	.66	.65	.65	.65	.64	.64	.63

MF
G-.65 ↑0
M-.55
P-.45 ↓79
Direct, RLM, 2 40-watt lamps, MS = 1.0 x MH

Room index	50%	30%	10%	50%	30%	10%	30%	10%
J	.38	.32	.28	.37	.32	.28	.31	.28
I	.47	.42	.39	.46	.41	.38	.40	.37
H	.51	.47	.44	.50	.47	.43	.46	.43
G	.55	.51	.48	.54	.51	.47	.50	.47
F	.58	.54	.51	.57	.53	.51	.52	.50
E	.63	.60	.57	.62	.59	.56	.58	.55
D	.68	.64	.61	.66	.64	.61	.63	.60
C	.70	.67	.63	.68	.65	.64	.64	.62
B	.73	.70	.68	.71	.68	.67	.67	.66
A	.74	.72	.70	.72	.70	.68	.69	.67

MF
G-.65 ↑0
M-.55
P-.45 ↓72
Direct, RLM, 3 40-watt lamps, MS = 1.0 x MH

Room index	50%	30%	10%	50%	30%	10%	30%	10%
J	.34	.29	.25	.33	.29	.25	.28	.25
I	.42	.38	.35	.41	.37	.34	.37	.34
H	.46	.42	.39	.44	.42	.39	.41	.39
G	.50	.46	.43	.48	.45	.41	.44	.41
F	.53	.49	.46	.51	.47	.44	.47	.44
E	.57	.54	.51	.56	.52	.50	.52	.50
D	.61	.58	.55	.59	.56	.54	.56	.54
C	.63	.60	.57	.61	.58	.56	.58	.56
B	.66	.64	.61	.64	.60	.59	.60	.59
A	.67	.65	.62	.66	.62	.61	.62	.61

MF
G-.60 ↑0
M-.50
P-.45 ↓71
Direct, RLM, 2·85-watt lamps, MS = 1.0 x MH

Room index	50%	30%	10%	50%	30%	10%	30%	10%
J	.33	.28	.25	.33	.28	.25	.28	.25
I	.41	.37	.34	.40	.36	.33	.36	.33
H	.45	.41	.38	.44	.41	.38	.40	.38
G	.48	.45	.42	.48	.45	.42	.43	.42
F	.51	.48	.45	.50	.47	.45	.46	.45
E	.55	.53	.50	.55	.52	.50	.51	.50
D	.60	.57	.54	.58	.56	.54	.55	.54
C	.61	.59	.56	.60	.57	.56	.57	.55
B	.64	.62	.60	.62	.60	.59	.60	.58
A	.65	.63	.61	.64	.62	.60	.61	.60

MF
G-.70 ↑0
M-.65
P-.55 ↓60
Direct, dust and vapor-tight, MS = 1.0 x MH

Room index	50%	30%	10%	50%	30%	10%	30%	10%
J	.29	.26	.23	.28	.26	.23	.25	.23
I	.35	.32	.31	.35	.32	.30	.32	.30
D	.38	.36	.34	.38	.36	.34	.35	.34
G	.41	.39	.37	.41	.39	.37	.38	.37
F	.44	.41	.39	.42	.41	.39	.40	.39
E	.46	.45	.42	.46	.44	.42	.44	.42
D	.50	.48	.46	.49	.47	.46	.46	.46
C	.51	.49	.47	.50	.48	.47	.48	.46
B	.53	.51	.50	.52	.50	.49	.49	.49
A	.54	.52	.50	.53	.51	.50	.50	.49

Table 2-6. Coefficients of Utilization (Continued)

Luminaire	Ceiling...	75%			50%			30%	
	Walls...	50%	30%	10%	50%	30%	10%	30%	10%
	Room index	Coefficient of utilization							
MF G−.70 ↑0 M−.60 P−.50 ↓80 Direct, 3-kw mercury, MS = 1.0 x MH	J	.38	.32	.28	.37	.32	.28	.31	.28
	I	.47	.42	.39	.46	.41	.38	.41	.38
	H	.51	.47	.43	.50	.47	.43	.46	.43
	G	.55	.51	.47	.54	.51	.47	.49	.47
	F	.58	.54	.51	.56	.53	.81	.52	.51
	E	.63	.59	.56	.62	.59	.56	.58	.56
	D	.67	.64	.61	.66	.63	.61	.63	.61
	C	.69	.67	.64	.67	.65	.63	.64	.63
	B	.72	.70	.67	.71	.68	.67	.67	.66
	A	.74	.71	.69	.72	.70	.68	.69	.67
MF G−.65 ↑0 M−.55 P−.45 ↓64 Direct, RLM with louvers, MS = .9 x MH	J	.33	.28	.26	.32	.28	.26	.28	.26
	I	.39	.36	.34	.39	.35	.34	.35	.34
	H	.43	.40	.38	.42	.40	.38	.39	.38
	G	.46	.43	.41	.45	.43	.41	.42	.41
	F	.48	.46	.43	.47	.45	.43	.45	.43
	E	.52	.50	.47	.51	.49	.47	.48	.47
	D	.55	.53	.51	.54	.52	.51	.52	.51
	C	.57	.55	.52	.54	.53	.52	.53	.52
	B	.59	.57	.56	.57	.56	.55	.55	.54
	A	.60	.58	.56	.59	.57	.56	.56	.55
MF G−.70 ↑0 M−.60 P−.50 ↓50 Direct, Troffer, glass, MS = 1.0 x MH	J	.28	.27	.26	.28	.27	.26	.28	.26
	I	.34	.33	.32	.34	.32	.32	.33	.31
	H	.36	.36	.36	.36	.36	.35	.35	.35
	G	.39	.38	.38	.38	.38	.37	.37	.36
	F	.41	.40	.39	.40	.39	.38	.39	.38
	E	.43	.42	.41	.42	.42	.40	.41	.40
	D	.46	.44	.43	.44	.43	.43	.42	.42
	C	.46	.45	.44	.45	.44	.43	.43	.43
	B	.47	.45	.45	.46	.44	.44	.44	.44
	A	.47	.46	.46	.46	.45	.45	.45	.44
MF G−.70 ↑0 M−.60 P−.50 ↓47 Direct, Troffer, glass, MS = 1.0 x MH	J	.27	.25	.24	.26	.25	.24	.26	.24
	I	.32	.31	.30	.31	.30	.30	.30	.29
	H	.34	.34	.33	.34	.33	.33	.33	.32
	G	.36	.35	.35	.36	.35	.35	.35	.34
	F	.39	.38	.37	.37	.37	.36	.37	.36
	E	.40	.40	.38	.39	.39	.38	.39	.38
	D	.43	.41	.40	.41	.40	.40	.40	.39
	C	.43	.42	.41	.42	.41	.40	.40	.40
	B	.44	.43	.42	.43	.42	.41	.41	.41
	A	.45	.44	.43	.43	.42	.42	.42	.41
MF G−.70 ↑0 M−.60 P−.55 ↓61 Direct, Troffer, louvers, MS = .8 x MH	J	.33	.31	.30	.33	.31	.30	.30	.29
	I	.40	.38	.38	.39	.38	.37	.38	.36
	H	.43	.42	.41	.42	.41	.41	.41	.40
	G	.46	.45	.44	.46	.44	.43	.44	.43
	F	.49	.47	.46	.47	.46	.45	.46	.45
	E	.51	.50	.49	.50	.49	.48	.49	.47
	D	.55	.52	.51	.53	.52	.50	.51	.50
	C	.55	.54	.52	.54	.53	.52	.52	.51
	B	.56	.55	.54	.55	.53	.53	.53	.52
	A	.57	.56	.55	.56	.55	.53	.54	.53

Table 2-6. Coefficients of Utilization (Continued)

Luminaire	Ceiling..	75%			50%			30%	
	Walls...	50%	30%	10%	50%	30%	10%	30%	10%
	Room index	Coefficient of utilization							
MF G-.75↑8 M-.65 P-.55↓50 — Semidirect, surface-mounted, MS = 1.0 x MH	J	.21	.17	.14	.20	.16	.14	.16	.14
	I	.26	.22	.20	.25	.21	.19	.21	.19
	H	.29	.25	.23	.28	.25	.22	.24	.22
	G	.32	.28	.25	.30	.27	.25	.26	.24
	F	.34	.30	.27	.33	.30	.27	.29	.27
	E	.38	.34	.31	.36	.33	.31	.32	.30
	D	.41	.37	.34	.39	.36	.34	.35	.33
	C	.42	.39	.36	.41	.38	.36	.37	.35
	B	.45	.42	.39	.42	.40	.39	.39	.38
	A	.47	.44	.41	.45	.42	.40	.41	.39
MF G-.75↑9 M-.65 P-.55↓55 — Semidirect, surface-mounted, MS = 1.0 x MH	J	.24	.20	.19	.23	.20	.17	.19	.17
	I	.30	.26	.23	.29	.25	.23	.25	.23
	H	.33	.29	.27	.32	.29	.26	.28	.26
	G	.36	.32	.30	.34	.32	.29	.30	.29
	F	.39	.35	.32	.37	.34	.31	.33	.31
	E	.42	.39	.35	.41	.38	.35	.36	.34
	D	.45	.42	.39	.44	.41	.38	.40	.38
	C	.47	.44	.41	.45	.42	.40	.41	.39
	B	.50	.47	.44	.48	.45	.43	.44	.42
	A	.52	.49	.46	.50	.47	.45	.45	.44
MF G-.75↑18 M-.65 P-.55↓53 — Semidirect, surface-mounted MS = 1.0 x MH	J	.23	.19	.17	.23	.18	.16	.17	.16
	I	.29	.25	.22	.28	.24	.21	.23	.21
	H	.32	.26	.25	.31	.28	.25	.26	.24
	G	.36	.32	.29	.34	.30	.27	.29	.26
	F	.40	.35	.31	.37	.33	.30	.31	.29
	E	.43	.39	.35	.41	.37	.34	.35	.32
	D	.47	.42	.39	.44	.40	.37	.38	.36
	C	.49	.45	.41	.46	.42	.39	.40	.38
	B	.52	.48	.45	.49	.45	.43	.43	.41
	A	.54	.51	.47	.51	.47	.45	.44	.43
MF G-.75↑24 M-.65 P-.55↓66 — Semidirect, surface-mounted, MS = 1.0 x MH	J	.29	.24	.22	.29	.23	.20	.22	.20
	I	.37	.32	.28	.36	.30	.27	.29	.27
	H	.41	.35	.32	.39	.34	.32	.33	.30
	G	.46	.41	.37	.43	.38	.34	.37	.33
	F	.51	.44	.39	.47	.42	.39	.39	.37
	E	.55	.49	.44	.52	.47	.43	.44	.41
	D	.60	.53	.49	.56	.51	.47	.48	.46
	C	.62	.57	.52	.58	.53	.50	.51	.48
	B	.66	.61	.57	.62	.57	.54	.54	.52
	A	.68	.64	.60	.65	.60	.57	.56	.55
MF G-.75↑39 M-.70 P-.65↓45 — General diffuse, enclosing globe, MS = 1.2 x MH	J	.24	.20	.16	.22	.18	.16	.17	.15
	I	.30	.25	.23	.27	.23	.21	.22	.19
	H	.33	.29	.26	.31	.27	.24	.25	.22
	G	.37	.33	.30	.34	.30	.27	.27	.25
	F	.41	.36	.32	.36	.33	.31	.31	.27
	E	.45	.41	.37	.41	.37	.33	.33	.30
	D	.49	.44	.40	.44	.40	.37	.36	.33
	C	.51	.47	.43	.46	.42	.39	.38	.35
	B	.55	.51	.47	.49	.45	.43	.40	.38
	A	.57	.53	.50	.51	.47	.45	.42	.40

Table 2-6. Coefficients of Utilization (Continued)

Luminaire	Ceiling..	75%			50%			30%	
	Walls...	50%	30%	10%	50%	30%	10%	30%	10%
	Room index	Coefficient of utilization							
MF G-.75↓30 M-.65 P-.55↓59 Semidirect, ceiling–mounted,* MS = 1.0 x MH	J	.30	.25	.21	.28	.23	.20	.22	.19
	I	.38	.33	.29	.35	.30	.27	.29	.26
	H	.42	.37	.35	.39	.35	.32	.33	.30
	G	.46	.41	.37	.42	.38	.35	.35	.33
	F	.50	.45	.41	.45	.41	.38	.38	.36
	E	.55	.50	.46	.50	.46	.43	.43	.40
	D	.60	.55	.51	.54	.50	.47	.47	.45
	C	.62	.58	.54	.56	.52	.50	.49	.47
	B	.66	.62	.59	.60	.56	.54	.52	.50
	A	.68	.65	.61	.62	.58	.56	.54	.52
MF G-.70↑19 M-.65 P-.60↓49 Semidirect, ceiling–mounted,* 2 or 4 lamps, MS=.9 x MH	J	.28	.25	.23	.23	.21	.19	.18	.16
	I	.34	.31	.29	.28	.26	.25	.22	.21
	H	.37	.34	.33	.31	.29	.28	.25	.24
	G	.41	.38	.36	.35	.32	.30	.27	.26
	F	.43	.41	.38	.36	.33	.32	.29	.27
	E	.46	.44	.42	.39	.37	.35	.31	.30
	D	.50	.47	.45	.41	.39	.37	.33	.32
	C	.52	.49	.46	.42	.40	.39	.34	.33
	B	.54	.51	.50	.44	.42	.41	.36	.35
	A	.56	.53	.51	.46	.43	.42	.37	.36
MF G-.70↑46 M-.65 P-.60↓33 Direct-indirect, suspension-mounted, 2 or 4 lamps, MS = 1.2 x MH	J	.26	.23	.20	.23	.21	.19	.19	.17
	I	.31	.28	.27	.28	.26	.24	.23	.20
	H	.35	.32	.30	.31	.28	.27	.26	.24
	G	.38	.35	.33	.34	.31	.30	.28	.27
	F	.41	.38	.35	.36	.34	.32	.30	.28
	E	.44	.42	.39	.39	.37	.35	.32	.31
	D	.48	.45	.42	.42	.39	.38	.34	.33
	C	.50	.49	.44	.43	.41	.39	.35	.34
	B	.53	.50	.48	.45	.43	.42	.37	.36
	A	.54	.52	.50	.47	.45	.43	.39	.37
MF G-.65↑20 M-.55 P-.50↓47 Semidirect, ceiling–mounted,* glass bottom, MS=.9 x MH	J	.28	.23	.21	.23	.20	.18	.17	.16
	I	.33	.30	.28	.28	.25	.23	.22	.20
	H	.36	.33	.31	.30	.28	.26	.24	.23
	G	.39	.36	.34	.33	.30	.29	.26	.25
	F	.42	.39	.37	.35	.32	.31	.28	.26
	E	.45	.42	.40	.38	.36	.34	.30	.29
	D	.48	.45	.43	.40	.38	.36	.32	.31
	C	.50	.47	.44	.41	.39	.38	.33	.32
	B	.53	.50	.48	.43	.41	.40	.35	.34
	A	.54	.52	.49	.45	.43	.41	.36	.35
MF G-.65↑49 M-.55 P-.50↓33 Direct-indirect, suspension-mounted, glass bottom, MS = 1.2 x MH	J	.27	.24	.22	.24	.22	.21	.21	.19
	I	.33	.30	.29	.29	.27	.26	.25	.23
	H	.36	.33	.32	.32	.30	.29	.28	.26
	G	.39	.37	.35	.36	.33	.32	.30	.28
	F	.43	.40	.37	.39	.35	.34	.31	.30
	E	.46	.43	.41	.41	.38	.37	.34	.32
	D	.50	.46	.44	.43	.41	.39	.36	.35
	C	.52	.49	.46	.45	.43	.41	.37	.36
	B	.55	.52	.50	.47	.45	.44	.38	.37
	A	.56	.54	.52	.49	.47	.45	.40	.38
MF G-.60↑56 M-.50 P-.40↓20 Semi-indirect, suspension-mounted, 2 or 4 lamps, MS = 1.2 x MH	J	.18	.14	.13	.14	.12	.10	.09	.08
	I	.22	.19	.17	.18	.15	.14	.12	.11
	H	.25	.22	.20	.20	.18	.16	.14	.13
	G	.28	.25	.22	.22	.20	.18	.16	.15
	F	.30	.27	.24	.24	.22	.20	.17	.16
	E	.34	.30	.28	.27	.24	.22	.19	.18
	D	.37	.33	.31	.29	.26	.25	.21	.20
	C	.39	.36	.33	.30	.28	.26	.22	.21
	B	.42	.39	.37	.32	.30	.28	.24	.23
	A	.44	.41	.39	.34	.32	.30	.25	.24

* Data based upon photometric curve run with false ceiling plate installed above luminaire in accordance with standard test procedure.

Table 2-6. Coefficients of Utilization (Continued)

Luminaire	Ceiling.. 75			50%			30%	
	Walls... 50%	30%	10%	50%	30%	10%	30%	10%
Room index	Coefficient of utilization							

MF G−.70 ↑79 M−.60 P−.50 ↓3 — Indirect, glass, plastic, or metal, MS = 1.2 x MH

Room index	50%	30%	10%	50%	30%	10%	30%	10%
J	.16	.13	.11	.12	.10	.08	.06	.05
I	.20	.16	.15	.15	.13	.11	.08	.07
H	.23	.20	.17	.17	.14	.13	.10	.08
G	.26	.23	.20	.20	.17	.15	.11	.10
F	.29	.26	.23	.22	.19	.17	.12	.11
E	.32	.29	.26	.24	.21	.19	.13	.12
D	.36	.32	.30	.26	.24	.22	.15	.14
C	.38	.35	.32	.28	.25	.24	.16	.15
B	.42	.39	.36	.30	.29	.27	.18	.17
A	.44	.41	.39	.33	.30	.29	.19	.18

MF G−.65 ↑85 M−.60 P−.55 ↓0 — Indirect, silvered bowl, MS = 1.2 x MH

Room index	50%	30%	10%	50%	30%	10%	30%	10%
J	.17	.14	.12	.13	.11	.09	.07	.06
I	.21	.17	.15	.16	.14	.12	.09	.08
H	.24	.21	.18	.18	.15	.14	.11	.09
G	.27	.24	.21	.21	.18		.12	.11
F	.30	.27	.23	.23	.20	.18	.13	.12
E	.33	.30	.27	.25	.22	.20	.14	.13
D	.37	.33	.31	.27	.25	.23	.16	.15
C	.39	.36	.33	.29	.26	.25	.17	.16
B	.43	.40	.37	.31	.30	.28	.19	.18
A	.45	.42	.40	.34	.31	.30	.20	.19

Typical luminaire	Estimated maintenance factors*	Distribution and max spacing	Room index	Ceiling 75%			50%			30%	
				Walls 50%	30%	10%	50%	30%	10%	30%	10%

Luminous ceiling using thin corrugated plastic diffuser having a reflectance of .40 and transmittance of .50
G−.65 ↓0 M−.65 P−.45 ↓68 — Direct

Room index	50%	30%	10%	
J	.22	.16	.12	Estimates based on calculations with cavity reflectance = 75%, cavity efficiency = 60%, apparent ceiling reflectance = 60%, floor reflectance = 14%
I	.27	.22	.19	
H	.33	.28	.24	
G	.38	.32	.29	
F	.41	.37	.33	
E	.46	.42	.39	
D	.49	.46	.43	
C	.52	.49	.46	
B	.55	.52	.50	
A	.57	.55	.53	

45° plastic louverall below 2-lamp 40-watt industrial type fluorescent units and bare lamps
G−.65 ↓0 M−.65 P−.55 ↓60 — Direct

Room index	with reflectors shallow cavity 75%			without reflectors shallow cavity 75%		
	50%	30%	10%	50%	30%	10%
J	.28	.25	.23	.25	.20	.19
I	.31	.29	.27	.29	.25	.23
H	.34	.32	.30	.32	.28	.26
G	.37	.35	.33	.35	.32	.30
F	.40	.37	.35	.38	.34	.32
E	.43	.41	.38	.41	.38	.36
D	.45	.43	.40	.43	.40	.39
C	.46	.44	.42	.45	.42	.41
B	.48	.45	.43	.47	.44	.43
A	.48	.46	.44	.48	.46	.44

45° white metal louverall
G−.70 ↓0 M−.65 P−.55 ↓50 — Direct

Room index	with reflectors shallow cavity 75%			without reflectors shallow cavity 75%		
	50%	30%	10%	50%	30%	10%
J	.23	.20	.19	.23	.19	.18
I	.27	.24	.22	.26	.23	.21
H	.30	.27	.25	.29	.26	.24
G	.32	.29	.28	.32	.29	.27
F	.34	.31	.30	.34	.31	.29
E	.36	.33	.32	.36	.33	.32
D	.38	.35	.34	.38	.35	.34
C	.39	.37	.36	.39	.37	.36
B	.41	.39	.38	.41	.38	.38
A	.42	.40	.39	.42	.40	.39

Table 2-6. Coefficients of Utilization (Continued)

Luminaire	Ceiling..	75%			50%			30%	
	Walls...	50%	30%	10%	50%	30%	10%	30%	10%
	Room index	Coefficient of utilization							

MF G-.70↑0 M-.60 P-.55↓53
Direct, Troffer, louvers, MS = .8 x MH

Room index	50%	30%	10%	50%	30%	10%	30%	10%
J	.29	.27	.26	.29	.27	.26	.27	.26
I	.35	.34	.33	.35	.33	.33	.33	.31
H	.38	.37	.36	.37	.36	.36	.36	.35
G	.40	.39	.39	.40	.39	.38	.38	.38
F	.43	.42	.40	.41	.40	.39	.40	.39
E	.45	.44	.43	.44	.43	.42	.43	.42
D	.48	.46	.45	.45	.45	.44	.45	.44
C	.48	.47	.45	.45	.46	.45	.45	.45
B	.50	.48	.47	.48	.47	.46	.46	.46
A	.50	.49	.48	.49	.48	.47	.48	.46

MF G-.70↑0 Metal Plastic↑0 M-.65 P-.55↓50 ↓59
Direct, louverall ceiling, shielded to 45°, cavity reflectance 75%

Coefficients for plastic louvers based on the use of bare lamps without reflectors.

Room index	METAL Cavity—75%			PLASTIC Cavity—75%		
J	.23	.20	.19	.25	.20	.19
I	.27	.24	.22	.29	.25	.23
H	.30	.27	.25	.32	.28	.26
G	.32	.29	.28	.35	.32	.30
F	.34	.31	.30	.38	.34	.32
E	.36	.33	.32	.41	.38	.36
D	.38	.35	.34	.43	.40	.39
C	.39	.37	.36	.45	.42	.41
B	.41	.39	.38	.47	.44	.43
A	.42	.40	.39	.48	.46	.45

MF G-.70↑0 M-.60 P-.55↓60
Direct, surface-mounted, 2 or 4 lamps, MS = 1.0 x MH

Room index	50%	30%	10%	50%	30%	10%	30%	10%
J	.29	.26	.23	.28	.26	.23	.25	.23
I	.35	.32	.31	.35	.32	.30	.32	.30
H	.38	.36	.34	.38	.36	.34	.35	.34
G	.41	.39	.37	.41	.39	.37	.38	.37
F	.44	.41	.39	.42	.41	.39	.40	.39
E	.46	.45	.42	.46	.44	.42	.44	.42
D	.50	.48	.46	.49	.47	.46	.46	.46
C	.51	.49	.47	.50	.48	.47	.48	.46
B	.53	.51	.50	.52	.50	.49	.49	.49
A	.54	.52	.50	.53	.51	.50	.50	.49

MF G-.70↑5 M-.65 P-.60↓47
Semidirect, surface-mounted, MS = 1.0 x MH

Room index	50%	30%	10%	50%	30%	10%	30%	10%
J	.26	.23	.22	.25	.23	.22	.23	.21
I	.31	.29	.28	.30	.28	.27	.28	.26
H	.34	.32	.31	.32	.31	.30	.30	.29
G	.36	.34	.34	.35	.33	.32	.33	.32
F	.38	.36	.35	.36	.35	.34	.35	.33
E	.40	.39	.37	.39	.38	.36	.37	.35
D	.43	.41	.39	.41	.40	.38	.39	.38
C	.45	.42	.40	.42	.41	.39	.40	.39
B	.46	.44	.42	.44	.42	.41	.41	.40
A	.46	.45	.43	.45	.43	.42	.42	.41

NOTES: *Symbols*
MF   maintenance factor
G   good maintenance factor
M   medium maintenance factor
P   poor maintenance factor
MS   maximum spacing
MH   mounting height
Distribution. Curves represent shape and not quantity of light. Fluorescent curves are taken normal to lamp axis.

↑ 48 per cent up
↓ 36 per cent down
48 + 36 = 84% luminaire efficiency

Table 2-7. Room Index

Ceiling height in feet for semi-indirect and indirect lighting				9, 9½	10-11½	12-13½	14-16½	17-20	21-24	25-30	31-36	37-50
Mounting height above floor in feet for direct and semidirect lighting		7, 7½	8, 8½	9, 9½	10-11½	12-13½	14-16½	17-20	21-24	25-30	31-36	37-50
Room width, ft	Room length, ft	Room index*										
9 (8½-9)	8-10	H	I	J	J							
	10-14	H	I	I	J							
	14-20	G	H	I	J	J						
	20-30	G	G	H	I	J	J					
	30-42	F	G	H	I	J	J	J				
	42-up	E	F	G	H	I	J	J				
10 (9½-10½)	10-14	G	H	I	J	J						
	14-20	G	H	I	J	J	J					
	20-30	F	G	H	I	J	J					
	30-42	F	G	G	H	I	J	J				
	42-60	E	F	G	H	I	J	J				
	60-up	E	F	F	G	H	I	J				
12 (11-12½)	10-14	G	H	I	I	J	J					
	14-20	F	G	H	I	J	J					
	20-30	F	G	G	H	I	J	J				
	30-42	E	F	G	H	I	J	J				
	42-60	E	F	F	G	H	I	J				
	60-up	E	E	F	G	H	I	J				
14 (13-15½)	14-20	F	G	H	H	I	J	J				
	20-30	E	F	G	H	I	J	J				
	30-42	E	F	F	G	H	I	J	J			
	42-60	E	F	F	F	G	H	I	J	J		
	60-90	D	E	E	F	G	H	I	J	J		
	90-up	D	E	E	F	F	G	H	I	J		
17 (16-18½)	14-20	E	F	G	H	I	J	J				
	20-30	E	E	F	G	H	H	I	J			
	30-42	D	E	F	G	H	H	I	J	J		
	42-60	D	E	E	F	G	G	I	J	J	J	
	60-110	D	E	E	F	G	G	I	J	J	J	J
	110-up	C	D	E	E	F	G	H	I	J	J	
20 (19-21½)	20-30	D	E	F	G	H	I	J	J			
	30-42	D	E	E	F	G	H	I	J	J		
	42-60	D	D	E	E	F	G	H	I	J	J	
	60-90	C	D	E	E	F	G	H	I	J	J	
	90-140	C	D	D	E	F	F	H	I	I	J	J
	140-up	C	D	D	E	F	F	H	H	I	J	J
24 (22-26)	20-30	D	E	E	F	G	H	I	J	J		
	30-42	C	D	D	E	F	G	G	I	J	J	
	42-60	C	D	D	E	F	G	H	I	J	J	J
	60-90	C	D	D	D	E	F	H	I	J	J	J
	90-140	C	C	C	D	E	E	G	H	I	J	J
	140-up	C	C	D	E	E	F	G	H	I	J	
30 (27-33)	30-42	C	D	D	E	F	G	H	I	J	J	
	42-60	C	C	C	D	D	F	F	H	I	J	J
	60-90	B	C	C	C	D	E	F	G	H	J	J
	90-140	B	C	C	C	D	E	F	F	G	H	I	J
	140-180	B	C	C	C	D	E	F	F	G	H	I	J
	180-up	B	C	C	D	E	F	G	H	I	J	

Table 2-7. Room Index (Continued)

Ceiling height in feet for semi-indirect and indirect lighting			9, 9½	10–11½	12–13½	14–16½	17–20	21–24	25–30	31–36	37–50		
Mounting height above floor in feet for direct and semidirect lighting		7, 7½	8, 8½	9, 9½	10–11½	12–13½	14–16½	17–20	21–24	25–30	31–36	37–50
Room width, ft	**Room length, ft**					Room index*						
36 (34–39)	30–42	B	C	D	E	F	F	H	I	I	J	
	42–60	B	C	D	D	E	F	G	H	I	J	J
	60–90	A	C	C	E	E	F	H	H	J	J	
	90–140	A	B	C	C	C	D	E	F	G	H	H
	140–200	A	B	C	C	C	D	E	F	F	G	H
	200–up	A	B	C	C	D	E	F	G	H	I	
42 (40–45)	42–60	A	B	C	C	E	F	G	H	I	I	J
	60–90	A	B	B	C	D	E	F	G	H	I	J
	90–140	A	B	B	C	D	D	E	F	G	H	I
	140–200	A	A	B	C	D	D	E	E	F	G	H
	200–up	A	A	B	C	D	D	E	F	G	I	
50 (46–55)	42–60	A	A	B	C	D	E	F	G	H	I	J
	60–90	A	A	B	C	C	D	E	F	F	G	J
	90–140	A	A	A	C	C	D	E	E	F	G	I
	140–200	A	A	A	C	C	D	E	E	F	G	I
	200–up	A	A	A	C	C	D	E	E	F	G	H
60 (56–67)	60–90	A	A	A	B	C	D	E	F	G	H	I
	90–140	A	A	A	B	C	C	D	E	F	G	H
	140–200	A	A	A	B	C	C	D	E	E	F	H
	200–up	A	A	A	B	C	C	D	E	E	F	h
75 (68–90)	60–90	A	A	A	A	B	C	D	E	F	G	I
	90–140	A	A	A	A	B	C	D	E	F	G	H
	140–200	A	A	A	A	B	B	C	D	E	F	G
	200–up	A	A	A	A	B	B	C	D	E	F	G

* "Room index" is the classification of a room according to its proportions; large and small rooms of the same proportion have the same index. Hence, for large rooms of dimensions greater than those shown, divide each dimension by the same number and use the index determined for the smaller room.

EXAMPLE: A room 200 by 600 by 40 ft would have the same room index as a room 50 by 150 by 10 ft.

SOURCE: "Essential Data for Lighting Design," p. 2, General Electric Company, Cleveland, Ohio, November, 1951.

Table 2-8.   Diffuse-reflection Factors

Color	Average reflection factor	Color	Average reflection factor
White	0.88	Medium:	
Very Light:		Blue green	0.54
Blue green	.76	Yellow	.65
Cream	.81	Buff	.63
Blue	.65	Gray	.61
Buff	.76	Dark:	
Gray	.83	Blue	.08
Light:		Yellow	.50
Blue green	.72	Brown	.10
Cream	.79	Gray	.25
Blue	.55	Green	.07
Buff	.70	Black	.03
Gray	.73	Wood Finishes:	
		Maple	.42
		Walnut	.16
		Mahogany	.12

SOURCE: "Illumination Design Data for Interiors," pp. 2, 6, Westinghouse Electric Corporation, Bloomfield, N.J.

Table 2-9.   Brightness Ratios

Area	Max Ratio
Between task and surroundings	3:1
Between task and remote surfaces (walls)	10:1
Between fixtures and adjacent surface	20:1
Anywhere in normal field of view	40:1

SOURCE: "IES Handbook," Sec. 9. Illuminating Engineering Society, New York, 1952.

Table 2-10. Approximate Ballast Loss per Lamp, Watts

Bulb	Watts	Starter switch no. or current, ma	Approximate ballast loss per lamp, watts					
			110–125 volt				220–250 volt, high PF	
			Single lamp		Two-lamp high PF			
			Low PF	High PF	Series	Lead lag	Single lamp	Two lamp
Preheat Lamps								
48" T-12	40	FS-4	8.5	11	...	7.8	8.5	9.3
60" T-17	90	FS-85	....	25	...	19.5	...	16
Rapid-start Lamps								
48" T-12	40	430	12	12	...	8.5		
Instant-start Lamps								
48" and 60"	40	415	....	20	11	12		
Slimline Lamps								
72" T-12	55	425	....	25	16	15		
	67	600	....	...	...	17.5		
96" T-12	74	425	....	29	16	17.5		
	95	600	....	...	...	29		
96" T-8	50	200	....	20	...	16		
	69	300	....	25	...	23		

source: "Illumination Design Data for Interiors," pp. 2, 6, Westinghouse Electric Corporation, Bloomfield, N.J.

Table 2-11. Brightness Ratios for Direct, Uniformly Diffusing, Indirect, and Luminous-ceiling Systems

$$A = \frac{\text{average wall brightness (midway between floor and ceiling)}}{\text{average illumination at work plane}}$$

Ceiling reflectance	0.80				0.70			0.50		
Wall reflectance...	0.80	0.50	0.30	0.10	0.50	0.30	0.10	0.50	0.30	0.10
Room coef*	Luminous Ceilings or Indirect Luminaires (Floor Reflectance: 0.30)									
0.0	0.520	0.325	0.195	0.0650	0.325	0.195	0.0650	0.325	0.195	0.0650
0.1	.536	.332	.198	.0657	.332	.198	.0657	.332	.198	.0656
0.2	.551	.340	.202	.0667	.340	.202	.0667	.340	.202	.0667
0.3	.567	.348	.206	.0680	.348	.206	.0680	.348	.206	.0680
0.4	.583	.357	.212	.0697	.357	.212	.0697	.357	.212	.0696
0.5	.598	.367	.218	.0717	.367	.218	.0717	.367	.218	.0716
0.7	.631	.389	.231	.0765	.388	.231	.0765	.388	.231	.0762
1.0	.681	.426	.256	.0856	.426	.256	.0856	.426	.256	.0851
	Luminous Ceilings or Indirect Luminaires (Floor Reflectance: 0.10)									
0.0	.440	.275	.165	.0550	.275	.165	.0550	.275	.165	.0550
0.1	.463	.288	.172	.0570	.287	.172	.0570	.288	.172	.0570
0.2	.486	.300	.179	.0592	.300	.179	.0592	.300	.179	.0592
0.3	.508	.313	.186	.0616	.313	.186	.0615	.313	.186	.0615
0.4	.530	.326	.194	.0641	.327	.194	.0641	.327	.194	.0641
0.5	.552	.340	.202	.0668	.340	.202	.0668	.340	.202	.0668
0.7	.594	.368	.220	.0728	.368	.220	.0728	.368	.220	.0728
1.0	.655	.413	.249	.0833	.413	.249	.0833	.413	.249	.0833
	Direct Luminaires (Floor Reflectance: 0.30)									
0.0	.218	.137	.082	.0273	.128	.077	.0255	.113	.068	.0225
0.1	.329	.193	.115	.0372	.189	.112	.3640	.182	.100	.0350
0.2	.342	.200	.115	.0367	.197	.113	.3610	.189	.109	.0350
0.3	.363	.205	.117	.0366	.202	.113	.3630	.194	.112	.0353
0.4	.384	.216	.119	.0370	.211	.118	.3670	.205	.115	.0361
0.5	.409	.224	.124	.0384	.220	.123	.3820	.217	.120	.0375
0.7	.415	.253	.139	.0429	.249	.139	.4280	.247	.137	.0424
1.0	.594	.331	.192	.0602	.328	.192	.6020	.327	.191	.0598
	Direct Luminaires (Floor Reflectance: 0.10)									
0.0	.072	.045	.027	.0090	.042	.026	.0085	.038	.023	.0075
0.1	.211	.127	.074	.0242	.125	.073	.0238	.062	.071	.0234
0.2	.239	.140	.080	.0255	.138	.079	.0253	.134	.077	.0249
0.3	.266	.151	.086	.0269	.150	.085	.0268	.146	.083	.0265
0.4	.298	.165	.092	.0287	.164	.089	.0286	.160	.090	.0282
0.5	.329	.181	.100	.0310	.179	.100	.0309	.176	.098	.0308
0.7	.402	.217	.121	.0372	.217	.120	.0372	.214	.119	.0370
1.0	.552	.308	.179	.0565	.307	.179	.0564	.304	.179	.0562
	Uniformly Diffusing Luminaires (Floor Reflectance: 0.30)									
0.0	.351	.220	.132	.0439	.209	.125	.0418	.183	.110	.0367
0.1	.696	.442	.268	.0903	.453	.274	.0925	.478	.289	.0977
0.2	.705	.455	.278	.0945	.464	.284	.0967	.490	.298	.1020
0.3	.714	.466	.287	.0987	.476	.294	.1009	.500	.306	.1063
0.4	.725	.478	.297	.1029	.488	.304	.1052	.512	.313	.1107
0.5	.736	.490	.307	.1074	.500	.314	.1096	.523	.320	.1151
0.7	.757	.514	.328	.1165	.523	.334	.1188	.545	.334	.1245
1.0	.790	.551	.359	.1308	.559	.364	.1329	.576	.354	.1377
	Uniformly Diffusing Luminaires (Floor Reflectance: 0.10)									
0.0	.596	.372	.223	.0744	.363	.218	.0727	.342	.205	.0683
0.1	.653	.415	.252	.0848	.424	.257	.0867	.447	.271	.0913
0.2	.663	.427	.269	.0853	.437	.268	.0912	.460	.382	.0960
0.3	.675	.441	.272	.0906	.450	.278	.0957	.473	.293	.1007
0.4	.688	.454	.283	.0959	.464	.289	.1003	.487	.304	.1055
0.5	.701	.468	.294	.1011	.477	.300	.1051	.500	.315	.1103
0.7	.729	.496	.316	.1115	.505	.322	.1148	.525	.336	.1201
1.0	.769	.538	.350	.1272	.545	.356	.1300	.562	.368	.1348

Table 2-11. Brightness Ratios for Direct, Uniformly Diffusing, Indirect, and Luminous-ceiling Systems (Continued)

$$B = \frac{\text{average ceiling brightness}}{\text{average illumination at work plane}}$$

Ceiling reflectance	0.80				0.70			0.50		
Wall reflectance...	0.80	0.50	0.30	0.10	0.50	0.30	0.10	0.50	0.30	0.10
Room coef*	Luminous Ceilings or Indirect Luminaires (Floor Reflectance: 0.30)									
0.0	1.000	1.000	1.000	1.0000	1.000	1.000	1.0000	1.000	1.000	1.0000
0.1	1.074	1.108	1.129	1.1510	1.108	1.129	1.1510	1.108	1.129	1.1510
0.2	1.153	1.228	1.277	1.3260	1.228	1.277	1.3260	1.228	1.277	1.3260
0.3	1.236	1.363	1.446	1.5280	1.363	1.446	1.5280	1.363	1.446	1.5280
0.4	1.324	1.514	1.638	1.7610	1.514	1.638	1.7610	1.514	1.638	1.7590
0.5	1.418	1.682	1.856	2.0300	1.682	1.856	2.0300	1.682	1.856	2.0270
0.7	1.625	2.078	2.386	2.6990	2.078	2.386	2.6990	2.078	2.386	2.6910
1.0	1.989	2.861	3.481	4.1380	2.861	3.481	4.1380	2.861	3.481	4.1120
	Luminous Ceilings or Indirect Luminaires (Floor Reflectance: 0.10)									
0.0	1.000	1.000	1.000	1.0000	1.000	1.000	1.0000	1.000	1.000	1.0000
0.1	1.086	1.114	1.134	1.1530	1.114	1.134	1.1520	1.114	1.134	1.1520
0.2	1.174	1.241	1.285	1.3280	1.241	1.285	1.3280	1.241	1.285	1.3280
0.3	1.267	1.381	1.457	1.5310	1.381	1.457	1.5310	1.381	1.457	1.5310
0.4	1.364	1.537	1.651	1.7650	1.537	1.651	1.7650	1.537	1.651	1.7650
0.5	1.466	1.710	1.872	2.0350	1.710	1.872	2.0350	1.710	1.872	2.0350
0.7	1.686	2.113	2.406	2.7050	2.113	2.406	2.7050	2.113	2.406	2.7050
1.0	2.067	2.906	3.506	4.1460	2.906	3.506	4.1460	2.905	3.506	4.1460
	Direct Luminaires (Floor Reflectance: 0.30)									
0.0	0.234	0.234	0.234	0.2340	0.210	0.210	0.2100	0.150	0.150	0.1500
0.1	.245	0.232	0.212	0.2020	0.196	0.188	0.1760	.140	.132	0.1260
0.2	.252	0.214	0.190	0.1700	0.186	0.166	0.1480	.133	.118	0.1060
0.3	.264	0.206	0.173	0.1440	0.179	0.151	0.1260	.127	.108	0.0897
0.4	.278	0.202	0.160	0.1230	0.175	0.140	0.1080	.124	.099	0.0769
0.5	.297	0.202	0.151	0.1070	0.175	0.131	0.0938	.124	.093	0.0668
0.7	.338	0.211	0.143	0.0847	0.183	0.125	0.0732	.129	.088	0.0528
1.0	.440	0.357	0.162	0.0737	0.224	0.143	0.0477	.158	.102	0.0460
	Direct Luminaires (Floor Reflectance: 0.10)									
0.0	.080	0.080	0.080	0.0800	0.070	0.070	0.0700	.050	.050	0.0500
0.1	.096	0.084	0.076	0.0689	0.073	0.067	0.0592	.052	.047	0.0430
0.2	.116	0.091	0.075	0.0602	0.079	0.065	0.0526	.056	.046	0.0386
0.3	.140	0.099	0.075	0.0534	0.085	0.065	0.0470	.061	.046	0.0334
0.5	.195	0.121	0.080	0.0451	0.105	0.070	0.0396	.074	.049	0.0285
0.7	.234	0.151	0.092	0.0419	0.131	0.080	0.0368	.094	.057	0.0262
1.0	.394	0.223	0.134	0.0489	0.194	0.117	0.0436	.136	.083	0.0310
	Uniformly Diffusing Luminaires (Floor Reflectance: 0.30)									
0.0	.578	0.578	0.578	0.5780	0.535	0.535	0.5350	.433	.433	0.4330
0.1	.637	0.634	0.633	0.6310	0.589	0.587	0.5860	.479	.477	0.4760
0.2	.689	0.694	0.693	0.6950	0.642	0.643	0.6450	.522	.520	0.5240
0.3	.735	0.748	0.758	0.7670	0.694	0.732	0.7620	.561	.563	0.5770
0.4	.777	0.805	0.826	0.8480	0.745	0.764	0.8490	.602	.606	0.6350
0.5	.817	0.864	0.898	0.9360	0.797	0.830	0.9420	.640	.650	0.6970
0.7	.889	0.982	1.050	1.1340	0.902	0.969	1.1400	.715	.740	0.8330
1.0	.988	1.167	1.310	1.4830	1.062	1.195	1.4850	.825	.878	1.0610
	Uniformly Diffusing Luminaires (Floor Reflectance: 0.10)									
0.0	.889	0.889	0.889	0.8890	0.824	0.824	0.8240	.667	.667	0.6670
0.1	.565	0.564	0.563	0.5610	0.523	0.522	0.5210	.425	.424	0.4230
0.2	.631	0.634	0.637	0.6120	0.589	0.591	0.5930	.478	.480	0.4810
0.3	.689	0.703	0.712	0.6990	0.651	0.660	0.6690	.527	.535	0.5430
0.4	.741	0.769	0.789	0.7920	0.711	0.731	0.7510	.574	.590	0.6070
0.5	.790	0.835	0.869	0.8910	0.771	0.803	0.8380	.619	.645	0.6750
0.7	.878	0.967	1.040	1.1000	0.888	0.953	1.0260	.704	.757	0.8190
1.0	.993	1.166	1.305	1.4650	1.061	1.190	1.3450	.824	.929	1.0550

Table 2-11. Brightness Ratios for Direct, Uniformly Diffusing, Indirect, and Luminous-ceiling Systems (Continued)

$$C = \frac{\text{average floor brightness}}{\text{average illumination at work plane}}$$

Ceiling reflectance	0.80				0.70			0.50		
Wall reflectance...	0.80	0.50	0.30	0.10	0.50	0.30	0.10	0.50	0.30	0.10
Room coef*										
Luminous Ceilings or Indirect Luminaires (Floor Reflectance: 0.30)										
0.0	0.300	0.300	0.300	0.3000	0.300	0.300	0.3000	0.300	0.300	0.3000
0.1	.293	.290	.288	.2860	.290	.288	.2860	.290	.288	.2860
0.2	.286	.280	.277	.2730	.280	.277	.2730	.280	.277	.2730
0.3	.279	.271	.266	.2610	.271	.266	.2610	.271	.266	.2610
0.4	.273	.262	.255	.2490	.262	.255	.2490	.262	.255	.2480
0.5	.266	.253	.245	.2370	.253	.245	.2370	.253	.245	.2370
0.7	.254	.236	.226	.2160	.236	.226	.2160	.236	.226	.2150
1.0	.237	.213	.200	.1880	.213	.200	.1880	.213	.200	.1860
Luminous Ceilings or Indirect Luminaires (Floor Reflectance: 0.10)										
0.0	.100	.100	.100	.1000	.100	.100	.1000	.100	.100	.1000
0.1	.097	.096	.096	.0954	.096	.096	.0954	.096	.096	.0954
0.2	.095	.093	.092	.0910	.093	.092	.0910	.093	.092	.0910
0.3	.092	.090	.088	.0868	.090	.088	.0868	.090	.088	.0868
0.4	.090	.087	.085	.0828	.087	.085	.0828	.087	.085	.0828
0.5	.087	.084	.081	.0790	.084	.081	.0789	.084	.081	.0789
0.7	.083	.078	.075	.0718	.078	.075	.0718	.078	.075	.0718
1.0	.077	.070	.066	.0623	.070	.066	.0623	.070	.066	.0623
Direct Luminaires (Floor Reflectance: 0.30)										
0.0	.300	.300	.300	.3000	.300	.300	.3000	.300	.300	.3000
0.1	.302	.300	.299	.2980	.301	.300	.2990	.301	.300	.2990
0.2	.302	.300	.299	.2990	.301	.300	.2980	.301	.299	.2980
0.3	.303	.300	.299	.2980	.301	.299	.2970	.302	.300	.2980
0.4	.303	.301	.298	.2970	.301	.299	.2970	.303	.300	.2980
0.5	.303	.302	.300	.2990	.302	.300	.2990	.303	.302	.2990
0.7	.303	.302	.301	.3000	.302	.301	.2990	.305	.303	.3000
1.0	.302	.302	.302	.3010	.303	.302	.3010	.306	.304	.3010
Direct Luminaires (Floor Reflectance: 0.10)										
0.0	.100	.100	.100	.1000	.100	.100	.1000	.100	.100	.1000
0.1	.101	.100	.100	.1000	.101	.100	.1000	.101	.100	.1000
0.2	.101	.101	.100	.1000	.101	.100	.1000	.101	.100	.1000
0.3	.101	.101	.100	.1000	.101	.100	.1000	.101	.100	.1000
0.4	.101	.101	.100	.1000	.101	.100	.1000	.101	.100	.1000
0.5	.101	.101	.100	.1000	.101	.100	.1000	.101	.100	.1000
0.7	.100	.100	.100	.1000	.101	.100	.1000	.101	.100	.1000
1.0	.100	.100	.100	.1000	.100	.100	.1000	.101	.101	.1000
Uniformly Diffusing Luminaires (Floor Reflectance: 0.30)										
0.0	.300	.300	.300	.3000	.300	.300	.3000	.300	.300	.9000
0.1	.300	.298	.295	.2930	.298	.296	.2940	.300	.297	.2950
0.2	.300	.295	.291	.2870	.296	.292	.2880	.299	.293	.2910
0.3	.298	.292	.287	.2820	.293	.289	.2840	.298	.289	.2870
0.4	.296	.289	.283	.2770	.291	.285	.2800	.296	.285	.2840
0.5	.293	.286	.280	.2740	.288	.282	.2760	.295	.281	.2810
0.7	.288	.281	.275	.2680	.284	.278	.2710	.292	.274	.2770
1.0	.280	.275	.270	.2630	.279	.274	.2660	.288	.265	.2740
Uniformly Diffusing Luminaires (Floor Reflectance: 0.10)										
0.0	.100	.100	.100	.1000	.100	.100	.1000	.100	.100	.1000
0.1	.100	.099	.099	.0980	.100	.099	.0982	.100	.099	.0986
0.2	.100	.098	.097	.0920	.099	.098	.0964	.100	.099	.0973
0.3	.099	.097	.096	.0914	.098	.096	.0948	.099	.098	.0960
0.4	.098	.096	.095	.0906	.097	.095	.0934	.099	.097	.0949
0.5	.097	.095	.093	.0898	.096	.094	.0922	.098	.096	.0940
0.7	.095	.093	.091	.0884	.094	.092	.0902	.097	.095	.0925
1.0	.092	.091	.090	.0871	.092	.091	.0887	.095	.094	.0914

* Room coef = $\dfrac{\text{room height} \times (\text{room length} + \text{width})}{2 \times \text{room length} \times \text{width}}$

SOURCE: "IES Handbook," p. 9-35. Illuminating Engineering Society, New York, 1952.

Visual comfort is determined by first calculating the glare factor, which is given by

$$\text{Glare factor} = AB^2/D^2\alpha^2S^{0.6} \qquad (2\text{-}10)$$

where A = apparent area of source, in.^{2}
$\quad\ B$ = brightness of source, fl, divided by 1,000
$\quad\ D$ = distance from source to eye, ft, divided by 10
$\quad\ \alpha$ = angle above horizontal, deg, divided by 10
$\quad\ s$ = surrounding brightness, fl, divided by 10

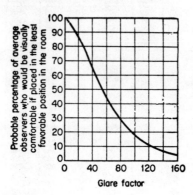

Fig. 2-1. The practical relationship between glare factor and the percentage of occupants who will be visually comfortable.

The total glare factor equals the sum of the glare factors for each fixture in a given reference direction. The total glare factor is entered in the curve of Fig. 2-1 to determine the visual comfort. The values in the equation are found from the physical size of the fixtures (area) and brightness curves. Correct lamp application for color is determined from Tables 2-12 and 2-13. Table 2-10 gives typical losses for fluorescent-lamp ballast, which serves as an aid in the determination of branch-circuit loading.

Table 2-12. Color of Lamps

Lamp type	Basic features	Use
Incandescent....	Oranges emphasized	General lighting, highlighting special areas, lighting meats and meat products
Standard cool white	Deficient in red	Displaying woods and furs, general lighting where the effect of color is not critical
De luxe cool white	Same appearance as cool white with red added	General lighting
Standard warm white	Beige tint emphasizes yellow greens	Home lighting (favorable to complexions)
De luxe warm white	Same as standard warm white with red added	Home lighting (favorable to complexions), use alone where warm red is needed in abundance
White..........	Emphasizes yellows, yellow greens, and oranges	Food-store lighting except for meats, general lighting of schools, offices, etc.
Daylight........	Emphasizes blues and greens, tends to gray red, oranges, and yellows	General lighting, displaying iced foods

Table 2-13 Effect of Colored Light on Colored Objects

Object color	Red light	Blue light	Green light	Yellow light
White.......	Light pink	Very light blue	Very light green	Very light yellow
Black.......	Reddish black	Blue black	Greenish black	Orange black
Red........	Brilliant red	Dark bluish red	Yellowish red	Bright red
Light blue...	Reddish blue	Bright blue	Greenish blue	Light reddish blue
Dark blue...	Dark reddish purple	Brilliant blue	Dark greenish blue	Light reddish purple
Green.......	Olive green	Green blue	Brilliant green	Yellow green
Yellow......	Red orange	Light reddish brown	Light greenish yellow	Brilliant light orange
Brown.......	Brown red	Bluish brown	Dark olive brown	Brownish orange

SOURCE: "Westinghouse Lighting Handbook," p. 4–16. Westinghouse Electric Corporation, Bloomfield, N.J., 1954.

INDEX

AC motors, 1-102
Air conditioning, electrical load for, 1-98
Air distribution, 1-68 to 1-76
   duct velocities, 1-70
   return outlets, 1-68
Air infiltration:
   through doors, 1-15
   through walls, 1-15
   through windows, 1-14
Air requirements:
   for building occupants, 1-17
   for odor removal, 1-16
Air volume, correction factors for altitude and temperature, 1-76
Appliances:
   gas, 1-91
   heat gain from, 1-63

Ballast loss for lamps, 2-20
Baseboard heaters, 1-18
Bays, interior, construction schemes for, 1-6, 1-7
Brightness ratio, 2-4, 2-19, 2-21 to 2-23
Building materials, weights of, 1-2
Burglar alarms, 1-107
Bus ducts, electrical, 1-101

Ceilings, weights of materials used for, 1-3
Centrifugal fans, 1-73
   performance curves of, 1-74 to 1-76
Chilled-water systems, 1-77
Chimneys, 1-24
Circuit breakers, 1-102
Coefficients of utilization, 2-9 to 2-16
Cold water for buildings, 1-59
Color of lamps, 2-25
Communication systems, 1-107
Condenser water systems, 1-77
Construction costs, 1-5
Convector ratings, 1-17
Conversion factors, illumination, 2-3
Cooling of buildings, 1-29 to 1-68
Costs for building construction, 1-5

Degree-days, number of, in heating season, 1-23
Design loads, 1-2
Design temperature, outdoor, 1-10
Diffuse reflection, 2-19
Drainage, 1-79
Drains and sewers, sizing, 1-82
Duct velocities, 1-70
Ducts, friction loss in, 1-71
Ductwork for air distribution, 1-70

Earth connections for grounding, 1-110
Electrical distribution, 1-99
Electrical loads in buildings, 1-68
Electrical power systems in buildings, 1-92

Fans, centrifugal, 1-73
Finned-tube heaters, 1-18
Fire alarms, 1-108
Fire retardant construction, 1-8
Flat roofs:
   heat gain through, 1-38 to 1-42
      temperature differentials for calculating, 1-38
   heat transfer coefficients of, 1-40 to 1-42
Floors, weights of materials used for, 1-3
Fluorescent lamps, 2-8
Framing systems, 1-4 to 1-9
Friction loss in ducts, 1-70

Gas piping in buildings, 1-91
Glare factor, 2-24
Grounding, electrical, for buildings, 1-109
Grounding low-voltage systems, 1-111

Heat-gain lighting fixtures, 1-30, 1-31
Heat gains:
   from appliances, 1-63
   effect of shading coefficients, 1-56 to 1-60
   from electric motors, 1-64

I-1

Heat gains (*Cont.*):
  through flat roofs, 1-38 to 1-42
  from people, 1-62
  through pitched roofs, 1-43
  solar factors for, 1-46 to 1-55
  through sunlit walls, 1-32 to 1-36
Heat transfer:
  for buildings, 1-12
  for doors, 1-12
  for flat roofs, 1-40 to 1-42
  for pitched roofs, 1-43
  for windows, 1-12, 1-61
Heating of buildings, 1-9 to 1-25
Heating season, length of, 1-23
Heating units, various types of, 1-18
Hot water for buildings, maximum daily
    requirements for, 1-88

Illumination, 2-2
  colored, 2-25
  room index for, 2-17
  [*See also* Lamp(s); Lighting]
Incandescent lamps, 2-7
Indoor design temperatures, 1-11
Insulation, effect of, on heat gain, 1-45

Kitchens, hot-water demand in, 1-89

Lamp(s):
  ballast loss per, 2-20
  color of, 2-25
  fluorescent, 2-8
  incandescent, 2-7
  mercury, 2-7
Lighting, 2-2
  coefficients of utilization, 2-9 to 2-16
  colored, 2-25
  fluorescent, 2-8
  incandescent, 2-7
  mercury, 2-7
  recommended levels of, 2-5
  room index for, 2-17
Lighting fixtures, heat gain from, 1-30,
    1-31
Lightning protection grounding, 1-110
Load(s):
  heating, 1-9 to 1-25
  live, 1-4
  structural, 1-2 to 1-4

Makeup water, 1-78
Material-handling, electrical loads, 1-99
Mercury lamps, 2-7
Moisture content in air, 1-26, 1-29

Motors, electrical, 1-102
  classification of, 1-104
  heat gain from, 1-64

Odors, removal of, 1-16
Outdoor air requirements, 1-17
Outdoor design temperatures, 1-10

Paging systems, 1-109
Panel boards, 1-102
Partitions, weights of material used for,
    1-3
Performance curves, centrifugal fans,
    1-74 to 1-76
Permeability of materials, 1-27
Pipe sizing for drains and sewers, 1-82
Pitched roofs, heat transfer coefficients
    for, 1-43
Plumbing fixtures, nonintegral traps for,
    1-80
Power systems, electrical, 1-92
Pressure losses in ducts, 1-70
Pressure switches, electrical, 1-102
Psychrometric chart, 1-69

Radiator ratings, 1-17
Reflection factors, 2-19
Refrigerators, heat gain from, 1-65
Return outlets for air distribution, 1-68
Roof gutters, sizing, 1-83
Roofing, weights of materials used for,
    1-3
Room index for lighting, 2-17
Runoff of storm water, 1-82

Safety switches, electrical, 1-102
Sanitary loads, 1-79
Sanitary vents, 1-81
Saturated air, vapor pressure of, 1-29
School alarm systems, 1-108
Sewers, 1-82
Shading coefficients, 1-56 to 1-60
Signal systems, 1-107
Smoke detection, 1-108
Snow loads on buildings, 1-4
Solar heat gain factors, 1-46 to 1-55
Stacks, 1-24
Steam flow rate and velocity, charts,
    1-19, 1-21
Storm drains, horizontal, 1-84
Storm water, 1-82
Sunlit walls, heat gain through,
    temperature differentials, 1-32 to
    1-36

Telephone zone closets, 1-106
Temperature:
    indoor, 1-11
    for inside design conditions, 1-65
    outdoor, 1-10, 1-66, 1-67

Vapor pressure of saturated air,
        1-29
Ventilating fans, 1-74
Ventilation load, 1-16
Vents, sanitary, 1-81
Voltage systems in buildings, 1-93

Walls, weights of material used for, 1-3
Water:
    cold, for buildings, 1-59
    consumption of, in buildings, 1-86
    distribution of, 1-77
    hot, for buildings, 1-88
    makeup, 1-78
    permeability of materials to, 1-27
Water cooling, 1-77
Weights of building members, 1-3
Wind loads used for buildings, 1-4
Windows, heat transfer coefficients for,
        1-61